Nuclear Radiation Detector Materials

ISSN 0272-9172

Volume 1—Laser and Electron-Beam Solid Interactions and Materials Processing, J.F. Gibbons, L.D. Hess, T.W. Sigmon, 1981

Volume 2—Defects in Semiconductors, J. Narayan, T.Y. Tan, 1981

Volume 3—Nuclear and Electron Resonance Spectroscopies Applied to Materials Science, E.N. Kaufmann, G.K. Shenoy, 1981

Volume 4—Laser and Electron-Beam Interactions with Solids, B.R. Appleton, G.K. Cellar, 1982

Volume 5—Grain Boundaries in Semiconductors, H.J. Leamy, G.E. Pike, C.H. Seager, 1982

Volume 6—Scientific Basis for Nuclear Waste Management, S.V. Topp, 1982

Volume 7—Metastable Materials Formation by Ion Implantation, S.T. Picraux, W.J. Choyke, 1982

Volume 8—Rapidly Solidified Amorphous and Crystalline Alloys, B.H. Kear, B.C. Giessen, M. Cohen, 1982

Volume 9—Materials Processing in the Reduced Gravity Environment of Space, G.E. Rindone, 1982

Volume 10—Thin Films and Interfaces, P.S. Ho, K.N. Tu, 1982

Volume 11—Scientific Basis for Nuclear Waste Management V, W. Lutze, 1982

Volume 12—In Situ Composites IV, F.D. Lemkey, H.E. Cline, M. McLean, 1982

Volume 13—Laser-Solid Interactions and Transient Thermal Processing of Materials, J. Narayan, W.L. Brown, R.A. Lemons, 1983

Volume 14—Defects in Semiconductors, S. Mahajan, J.W. Corbett, 1983

Volume 15—Scientific Basis for Nuclear Waste Management VI, D.G. Brookins, 1983

Volume 16—Nuclear Radiation Detector Materials, E.E. Haller, H.W. Kraner, W.A. Higinbotham, 1983

Volume 17—Laser Diagnostics and Photochemical Processing for Semiconductor Devices, R.M. Osgood, S.R.J. Brueck, H.R. Schlossberg, 1983

Volume 18—Interfaces and Contacts, R. Ludeke, K. Rose, 1983

Volume 19—Alloy Phase Diagrams, L.H. Bennett, T.B. Massalski, B.C. Giessen, 1983

Volume 20—Intercalated Graphite, M.S. Dresselhaus, G. Dresselhaus, J.E. Fischer, M.J. Moran, 1983

Volume 21—Phase Transformation in Solids, T. Tsakalakos, 1983

MATERIALS RESEARCH SOCIETY SYMPOSIA PROCEEDINGS VOLUME 16

Nuclear Radiation Detector Materials

Symposium held November 1982 in Boston, Massachusetts, U.S.A.

EDITORS:

E.E. Haller
Department of Materials Science and Mineral Engineering and Lawrence Berkeley Laboratory, University of California, Berkeley, California, U.S.A.

H.W. Kraner
Instrumentation Division and Department of Physics, Brookhaven National Laboratory, Upton, New York, U.S.A.

and

W.A. Higinbotham
Department of Nuclear Energy, Brookhaven National Laboratory, Upton, New York, U.S.A.

NORTH-HOLLAND
NEW YORK · AMSTERDAM · OXFORD

Published by:

Elsevier Science Publishing Co., Inc.
52 Vanderbilt Avenue, New York, NY 10017

Sole distributors outside the USA and Canada:

Elsevier Science Publishers B.V.
P.O. Box 211, 1000 AE Amsterdam, The Netherlands

Library of Congress Cataloging in Publication Data

Main entry under title:

Nuclear radiation detector materials.

(Materials Research Society symposia proceedings,
ISSN 0272-9172; v. 16)

Includes indexes.
1. Nuclear counters—Materials—Congresses.
I. Haller, E.E. II. Kraner, H.W.
III. Higinbotham, William A. IV. Series.
QC787.C6N8 1983 539.7′7 83-8915
ISBN 0-444-00787-3 (Elsevier)

Manufactured in the United States of America

Contents

Preface

The 1982 Materials Research Society Annual Meeting provided a timely opportunity to organize a Symposium on NUCLEAR RADIATION DETECTOR MATERIALS. The previous and perhaps only meeting devoted specifically to detector materials occurred in 1966 and few publications exist that comprehensively describe the progress and the state of this field. The unique combination of properties of solid state nuclear radiation detectors - excellent energy resolution, high detection efficiency, good time resolution, and spatial resolution reaching the tens of micrometer range - have led to an explosive increase of their use in the physical sciences, in biology and medicine, in mineral exploration and as radiation monitors in the laboratory, in power plants and in a wide variety of field applications. These unique properties are due in large part to the very high quality of the materials which are used to fabricate such detectors. The importance of materials research for the field of nuclear radiation detectors can hardly be overestimated.

The main goal of this meeting was to bring together scientists and engineers working with a variety of detector materials and have them exchange results and ideas, discuss novel methods and techniques and encourage them to make use of the unique opportunity to interact with the attendees of the other symposia of the Annual Meeting. These symposia represented a large part of the extremely wide materials science horizon. Judging from our observations and from the comments of many participants, we conclude that we have reached our goal.

Several conclusions might be drawn from the work presented in this volume:

1) Great progress has been made in the scientific understanding of the elemental semiconductors, silicon and germanium, which in turn has helped in the development of detector-grade materials. The extreme requirements of purity, crystallographic quality and minority carrier lifetime continue to pose serious limitations on the production yield of large single crystals.

2) High-purity germanium has completely replaced doped materials which have been used for the past twenty years for the fabrication of lithium-drifted germanium (Ge(Li)) detectors.

3) Bismuth germanate (BGO), a scintillator with several attractive properties, is rapidly finding new applications and may soon compete with thallium-activated sodium iodide, the proven "workhorse" of scintillation detectors. It is exciting to consider the opportunities for basic materials research of this and other forthcoming novel scintillator materials.

4) Mercuric iodide and cadmium telluride continue to be developed for new and important applications. Steady progress is being made in the growth and characterization of single crystals of these "difficult" materials.

The editors hope that this volume will be useful to a broad range of engineers and scientists who wish to understand semiconductor and scintillation detectors and detector materials. We foresee that the progress in this field will demand a revision of this compendium in the near future.

February 8, 1983

E. E. Haller
W. A. Higinbotham
H. W. Kraner

Acknowledgments

The Organizers of the Symposium on NUCLEAR RADIATION DETECTOR MATERIALS for the 1982 Annual Meeting of the Materials Research Society wish to thank this Society for the opportunity to bring together a group of people working with or simply interested in semiconductor and scintillator materials and detectors. The competent and untiring efforts of the MRS Program Chairman, Elton Kaufmann and the Treasurer, Kathleen Taylor have been most appreciated.

The Symposium and its proceedings became a reality with the generous financial help of both governmental and corporate sponsors. We most sincerely appreciate the support of the Department of Energy, Office of Health and Environmental Research, represented by Drs. R. W. Wood and G. Goldstein. Their support not only for this symposium but also in long term commitment to detector materials research and detector development should be widely recognized and acclaimed. The contributions of the following corporate sponsors is greatly appreciated: Canberra Industries, Mr. Orren K. Tench; Harshaw Chemical Company, Dr. M. R. Farukhi; and EG&G Ortec Inc., Mr. Sanford Wagner. The Nuclear and Plasma Science Society of the I.E.E.E. is acknowledged for having endorsed and publicized the symposium.

It is a pleasure to thank and acknowledge the Invited Speakers and Session Chairmen for their efforts and dedication to an excellent meeting. The invited speakers include: L. S. Darken, Jr., M. R. Farukhi, P. A. Glasow, W. L. Hansen, G. S. Hubbard, O. H. Nestor, P. Siffert, L. van den Berg, and J. T. Walton. In addition to the organizers, R. N. Hall and P. Siffert also served as session chairmen and we appreciated their help.

The assistance of Barbara Gaer (BNL), Vicki Donelson (LBL) and Lynne Dory (LBL) with several organizational tasks is very much appreciated. The enormous help and dedication of Mrs. Marie Kalousek (BNL) in preparing this volume and in organizing the manuscripts is most thankfully acknowledged. Her efforts were crucial to this publication.

February 7, 1982

E. E. Haller
W. A. Higinbotham
H. W. Kraner

Nuclear Radiation Detector Materials

MATERIAL GROWTH
AND
CHARACTERIZATION

HIGH-PURITY GERMANIUM CRYSTAL GROWING

W. L. HANSEN[1] and E.E. HALLER[1,2]
[1]Lawrence Berkeley Laboratory and [2]Department of Materials Science
University of California, Berkeley, CA 94720 U.S.A.

ABSTRACT

The germanium crystals used for the fabrication of nuclear radiation detectors are required to have a purity and crystalline perfection which is unsurpassed by any other solid material. These crystals should not have a net electrically active impurity concentration greater than $10^{10}cm^{-3}$ and be essentially free of charge trapping defects.

Such perfect crystals of germanium can be grown only because of the highly favorable chemical and physical properties of this element. However, ten years of laboratory scale and commercial experience has still not made the production of such crystals routine. The origin and control of many impurities and electrically active defect complexes is now fairly well understood but regular production is often interrupted for long periods due to the difficulty of achieving the required high purity or to charge trapping in detectors made from crystals seemingly grown under the required conditions.

The compromises involved in the selection of zone refining and crystal grower parts and ambients is discussed and the difficulty in controlling the purity of key elements in the process is emphasized. The consequences of growing in a hydrogen ambient are discussed in detail and it is shown how complexes of neutral defects produce electrically active centers.

INTRODUCTION

High-purity germanium was developed as a material for fabricating nuclear radiation detectors to be used principally as high resolution γ-ray detectors. A semiconductor nuclear radiation detector [1] is a reverse biased p-n junction with a region depleted of mobile charges to a width sufficient to stop a significant fraction of the radiation of interest. For the case of a detector made of germanium and for the detection of nuclear γ-rays (MeV energy range), this width must be of the order of centimeters. The width of the depleted region is related to the net electrically active impurity concentration by the expression:

$$W^2 = \frac{2\varepsilon V}{q|N_A - N_D|} \tag{1}$$

where W is the width of the depleted region in cm, ε is the dielectric constant of the semiconductor ($\varepsilon = \varepsilon_r\varepsilon_0$; $\varepsilon_r = 16$ for Ge, $\varepsilon_0 = 8.85 \times 10^{-4} Fm^{-1}$), q is the unit electron charge in $(1.6 \times 10^{-19}$ C), V is the applied reverse voltage and N_A and N_D are the ionized acceptor and donor impurity concentrations per cm^3. For V = 1140 volts and W = 1cm,

$$|N_A - N_D| = 2 \times 10^{10}cm^{-3}. \tag{2}$$

Purity in the range of $10^{10}cm^{-3}$ was not thought to be attainable and for a number of years Ge detectors were made only by using lithium ion drift compensation [2] of $10^{14}cm^{-3}$ p-type doped crystals. Dissatisfaction with the ion drift compensation technique led to a proposal by Hall [3] in 1966 that attempts should be made to achieve the required purity directly. Less than four years after Hall's proposal, crystals of the requisite purity were being grown.

The procedure which has been developed for growing crystals of high-purity germanium consists of zone refining the polycrystal in carbon or silica smoked silica boats [4] and subsequently growing single crystals by the Czochralski [7] method from synthetic silica crucibles in a hydrogen atmosphere [5,6]. All commercial high-purity germanium single crystals are produced by this method because it is a good compromise between process complexity, purity and cost and gives a moderate yield of acceptable product. However, this process is not optimium for the production of high purity material in a chemical sense and it involves the acceptance of fairly high concentrations of silicon, oxygen and hydrogen [17]. These impurities can combine to produce electrically active centers or charge traps [12].

Single crystals of germanium cannot usefully be grown by the float zone technique which has been so successfully applied to silicon because the ratio of the melt surface tension to density is too small to support a floating zone of more than 1cm diameter. Initial attempts to grow detector grade germanium by the horizontal boat technique were abandoned because of difficulties encountered in achieving high purity; experimenters objected to the trapezoidal shape of detectors thus grown. The ability to grow crystals with circular cross-sections and arbitrarily selected diameters were among the important factors in dictating that all of the development of high-purity germanium crystal growth be done using the Czochralski technique.

On presenting the current state-of-the-art of high-purity germanium crystal growth, the crystal growers' point of view will dominate. A discussion of the growth parameters which affect crystalline perfection will be followed by a discussion of typical impurity distributions and the kinds of information which can be abstracted from these distributions. The origin and control of the important impurities found in high-purity crystals introduces the problem of aluminum removal. It is next shown how growing out of silica in a hydrogen atmosphere leads to significant concentrations of neutral impurities, complexes and possibly precipitates. The stability of precipitates and their interaction with aluminun is examined using thermochemistry. The discussion section leads to a suggestion for improving the method of growing high-purity germanium crystal in the future.

CRYSTAL GROWTH

Crystal perfection and defect thermodynamics

Crystal perfection is of much greater importance for nuclear radiation detectors than for other semiconductor devices. Transistors, diodes, etc., have electrically functional regions with volumes often less than $10^{-9}cm^3$ whose operation is controlled by added chemical impurities. In radiation detectors, chemical impurities are minimized so as to produce large sensitive regions and the functional volume of the device can exceed $100cm^3$. Therefore, charge trapping and recombination effects due to native defects (dislocations, vacancies, interstitials) which would be unnoticed in other types of devices, can be the dominant cause limiting the performance of detectors.

It is known from thermodynamics that it is not possible at any finite temperature to grow a crystal free of defects [8]. A crystal in equilibrium with its melt is in a state of minimum free energy F with respect to the concentration of native defects, so that:

$$F = E - TS, \quad (3)$$

where E is the energy needed to create a defect and S is the entropy gained from the defect. Restricting the discussion to vacancies, for n vacancies in a crystal of N atoms, the entropy gained by the generation of vacancies is the logarithm of the number of combinations of ways of removing the n atoms so that:

$$F = E - TS = nE_v - kT \log \frac{N!}{(N - n)!n!} \quad (4)$$

where E_v is the energy needed to create a vacancy. If the factorials are simplified by Stirling's approximation ($\log N! = N \log N - N$) and knowing that at equilibrium $(\partial F/\partial n)_T = 0$, we have for $n \ll N$:

$$n = N \exp(-E_v/kT). \quad (5)$$

For germanium at its melting point, $E_v \simeq 2$eV and T = 1200K,

$$n \simeq 10^{14}\text{cm}^{-3}. \quad (6)$$

Using the appropriate energy of formation, the same basic argument holds for interstitials and for vacancy-interstitial pairs (Frenkel defects) [8]. Because germanium crystals have very strong, covalent molecular bonds, the native defect concentrations are extremely low compared with metals, for example, where $n_v = 10^{17}$ to 10^{19}cm^{-3} [31].

Because of the exponential dependence of native defect concentration on temperature, a cooling crystal becomes highly supersaturated in defects which tends to condense on lower energy state sites such as impurities, the surface, dislocations or on each other. Vacancies and interstitials tend to annihilate each other and the remaining dominant species is the result of complicated precipitation kinetics and is different for each semiconductor. In high-purity dislocation-free germanium, the remaining species is vacancy precipitates (voids) [9,18]. (It might be noted that in silicon it is interstitials (swirl defects) [10]). So much energy is required to nucleate dislocations that they cannot form in thermal equilibrium--this is the reason why dislocation-free crystals can be grown. Dislocations, if present, can consume an unlimited number of vacancies and interstitials by translation or climb.

Dislocation-free high-purity germanium grown in a hydrogen ambient has turned out to be unsuitable for detector fabrication because of a deep level at $E_v + 0.072$eV with a concentration of about 10^{11}cm^{-3} [11,18]. This center has been identified as a divacancy-hydrogen complex (V_2H) [12]. Experience has shown that if the crystal contains at least 100 uniformly distributed dislocations cm^{-2}, the V_2H trap concentration will be too low to influence radiation detector properties. It has further been shown that if the local density of dislocations exceeds 10^4cm^{-2}, the dislocations themselves begin to act as charge trapping centers [13]. Meeting the requirement that the dislocation density must be everywhere between 10^2 and 10^4cm^{-2} over volumes that can exceed 100cm^3 has proved to be the greatest challenge to the art of crystal growing.

A simplified schematic of apparatus typically used to grow high-purity germanium crystals is shown in Fig. 1. The design goal in a Czochralski growth apparatus is to establish a one-dimensional thermal gradient so that finite growth will occur and, at the same time, to minimize radial gradients which give rise to thermal stresses in the growing crystal. In Czochralski growth, thermal stresses have their origin in differential thermal expansion that the growing crystal may encounter as the newly formed solid moves away from the melt-solid interface. That is, away from the melt boundary the crystal experiences a three-dimensional thermal gradient instead of the one-dimensional gradient that exists at the boundary. Thermal stress provides the only means of accumulating enough energy to launch a dislocation, since the thermal energy at the melting point is insufficient.

A crystal growing from a melt can never be completely free of thermal stress because this implies zero thermal gradient and, therefore, zero growth rate. The best that can be done is to arrange the thermal environment so that an accepable growth rate is achieved while keeping the stress in the grown crystal below the level required to multiply a dislocation. One criterion for low thermal stress is that the growing interface be flat and normal to the growth direction. A flat interface can be approximated by arranging the thermal environment so that when the crystal is growing slowly at full diameter, the melt-solid interface is convex when viewed from the melt. The pulling rate is then increased until the heat of crystallization balances the radial heat loss from radiation and convection. Because Czochralski growth is a dynamic process, there is only one "right" set of conditions for each part of the crystal and the axial and radial gradients are locked to a precise growth rate.

To initiate crystal growth, a low dislocation density seed is selected and it is "necked down" [14] by fast pulling until only very few dislocations remain. The diameter is then increased under conditions of moderate thermal stress until the proper number of dislocations is obtained. A great deal of experience is required to successfully achieve these conditions. Flat interface growth is maintained for the bulk of the crystal but the interface shape near the end of the crystal is generally neglected as this portion will be discarded. A typical evolution of the melt-solid interface shape during Czochralski growth is illustrated in Figure 2.

The proper control of dislocation density and distribution is also sensitive to crystallographic growth direction. Diamond lattice crystals such as germanium grow isotropically except for the (111) surface which is growing more slowly. Because of this, (111) crystals tend to develop a "facet" [15] which causes a flat interface growth even under conditions of considerable isothermal curvature and, as a result, they may have the thermal stress frozen in. For any other growth axis, interface curvature will lead to stresses which cause dislocation multiplication which has the effect of "decorating" the stress distribution. Pure edge dislocations in crystals grown along the (100) direction tend to line up with the crystal growth axis and can often be followed through the entire length of the crystal. This propagation habit makes dislocation counting very reliable in (100) crystals because chemical etches preferentially decorate dislocations which intersect the surface nearly normally. For all other growth directions, the dislocations propagate by kinks and jogs so that only a small fraction meet any surface in a way that leads to chemical decoration. The (113) growth direction has proven to be useful in that the dislocations are much less effective as charge traps and thus greater dislocation densities can be tolerated [16].

The preferred cross-sectional shape of detector crystals is circular. However, this can only be achieved with a strong axial thermal gradient, otherwise the shape will tend to be dictated by the isothermal condition: triangular for (111), square for (100) and rectangular for (110) growth axes. Once a growing crystal acquires pronounced side facets, growth stability is threatened because the facets increase thermal radiation due to increased surface area which, in turn, enhances facet growth and may lead to dendritic growth. All of these restraints become more severe as larger cross section crystals are grown. For particular crystal grower designs, the thermal environment can lead to conflicting demands such as increased axial gradient to improve growth stability and decreased axial gradient so that some combination of radial gradient and growth rate can produce a flat interface. This can happen, for example, when after-heaters are used to lower radial gradients; the condition of flat interface may need a melting rather than a growing crystal.

It is unreasonable to expect that the production of the required high quality crystals will ever become routine, especially when the demands of chemical purity are added.

Net impurity profiles

After the above review of basic intrinsic defects and before going into the details of impurity chemistry and purification, let us examine "typical" high-purity germanium crystals which have been grown under typical conditions.

An impurity profile is a plot of the net electrically active impurity concentration $|N_A - N_D|$ against % of the melt crystallized. The measurement of $|N_A - N_D|$ is performed at 77K because at room temperature the generation of electron-hole pairs across the band gap exceeds $|N_A - N_D|$ by several orders of magnitude in typical radiation detector crystals. At 77K the free carrier concentration will be very nearly $|N_A - N_D|$.

The $|N_A - N_D|$ profile is found by soldering wires to the ends of the crystal with indium, immersing it in liquid nitrogen, passing a constant current through it (e.g. 100μA) and recording the voltage drop at predetermined intervals. At high purities there is negligible impurity scattering so that the carrier mobility μ is a constant with constant temperature. For germanium, the mobility of both electrons and holes is 44000 cm^2/Vs at 77K. The measured resistivity ρ and the net impurity concentration are related by:

$$\rho = \frac{V}{I} \cdot \frac{A}{l} = \frac{1}{|N_A - N_D|\, e\mu}$$

where e = charge of the electron (= 1.6×10^{-19}C), A is the cross-sectional area of the crystal and l = length of the interval between voltage contacts (in our case 1cm). We find:

$$|N_A - N_D| = \frac{I}{e\mu V} \frac{l}{A} \tag{7}$$

Experience with a particular crystal growing furnace has shown that a casual inspection of the impurity profile can usually indicate the identity and concentration of the dominant impurities and often reveals defect complexes and radial impurity gradients. As a result of many analyses by photothermal ionization spectroscopy (PTIS) [17], deep level transient spectroscopy (DLTS) [17] and low temperature Hall effect [17], which indicate the identity and distribution of the impurities it has been found that the important ones have a characteristic signature in the impurity profile. The impurity profiles are a reflection of the real impurity segregation during growth.

The impurity segregation coefficient K is the ratio of the solubility of the impurity in the solid to the solubility in the melt and unless otherwise stated, refers to the equilibrium value. As an example of a signature in an impurity profile, for a single impurity with $K < 1$, the profile will show a constantly increasing impurity concentration from seed end to tail due to accumulation of impurity in the remaining melt.

Typical impurity profiles for germanium crystals grown in hydrogen out of silica crucibles are shown schematically in Fig. 3. The most common profile is that of Fig. 3a. It is dominated by non-segregating aluminum acceptors and segregating phosphorus donors. Figure 3b illustrates boron contamination; boron is the only electrically active element in germanium with a segregation coefficient greater than one. Figure 3c shows a dislocation-free crystal with divacancy-hydrogen acceptors (V_2H) dominating the profile. Under our crystal growing conditions, the V_2H concentration lies always between 1 and $3 \times 10^{11} cm^{-3}$. Figure 3d is the profile of a crystal which is mostly dislocation free at the seed end and is, therefore, dominated by (V_2H) acceptors in the first part of the crystal only.

Whenever the crystal has a radial impurity gradient, the test current is inhomogeneous and the impurity profile is distorted. Hall effect measurements by the Van der Pauw method on slices cut from crystals with so-called "coring" yield greatly reduced average Hall mobilities when the free carrier concentration at the periphery is much larger than at the center. The signature of radial impurity gradients or "coring" is shown in Fig. 3e where the dotted lines show more distortion for greater degrees of coring. These coring profiles are sometimes indicative of copper contamination.

The impurity profiles show several general characteristics: 1) they are dominated by the residual chemical impurities B, Al, P and by the complex V_2H; 2) under normal growing conditions, i.e., no air leaks, B and P segregate normally and Al and V_2H do not segregate; and 3) condition 2 automatically leads to the observation that whenever the crystal contains a p-n junction, it is always p-type at the head and n-type at the tail, never the reverse.

Purification and impurity control

Elemental impurities. Aluminum, an acceptor, is by far the most troublesome impurity to remove because it does not seem to segregate. It appears that the aluminum is bound to a dispersed stable phase which survives both zone refining and crystal growth and that the electrically active Al concentration is a result of the equilibrium constant for the reaction:

$$Al\text{-}X \leftrightarrows Al_{(1)} + X_{(1)} \quad (8)$$

where X is almost certainly SiO_2. This point will be discussed again later.

Phosphorus, a donor, is a ubiquitous impurity which is best reduced by good housekeeping and particulate control. Phosphorus does not form compounds that are stable at the melting point of germanium and it is readily removed by zone refining. Experience suggests that synthetic silica crucibles are the principal source of phosphorus. Because it segregates normally during crystal growth, phosphorus is not usually the dominant cause for rejection of a crystal, at least not for the first half grown.

Boron, an acceptor, forms a very stable oxide and if its presence is suspected, it can be removed by oxidation and precipitation of the oxide rather than by segregation. One zone pass under slightly oxidizing conditions [19] removes boron effectively. The segregation coefficient for boron is >1 so that it is not easily removed by zone refining.

Copper behaves in a peculiar way in germanium. On the one hand, it has a very small segregation coefficient (10^{-5}) and very low solubility so that it should be easily removed, but on the other hand, it has a very high diffusion coefficient at relatively low temperatures so that it is easy to recontaminate the crystal. Copper is a deep level multiple acceptor in germanium and is a strong hole trap at 77K.

One signature of copper contamination in germanium can be a radial impurity gradient. The small segregation coefficient of copper seems to make it almost impossible to contaminate the crystal through the melt. It is more likely that the crystal is contaminated through the gas phase. Measurements show the copper distribution in the crystal is strongly influenced by the time-temperature history of the growing crystal. Figure 4 illustrates this kind of distribution in a copper contaminated crystal which would be otherwise high-purity n-type. The resultant distribution is described as "coring" It depends on where in the crystal the sample is cut and on the underlying shallow impurity concentration and type.

Copper forms many high vapor pressure, low stability compounds [CuOH, $Cu(OH)_2$, $Cu_2(OH)_3Cl$, $CuCl_2 \cdot 3CuO \cdot 4H_2O$, etc.] and care must be exercised in excluding it from the crystal grower environment. We have found that rubber gaskets can be an important copper source. Problems with copper contamination are variable from facility to facility. In our laboratory, copper is not normally present in the as-grown cyrstals when measured with a sensitivity of $\leq 10^8 cm^{-3}$ (DLTS). When copper contamination tends to occur, it can usually be suppressed by cleaning the crystal grower parts.

Neutral impurities, complexes and precipitates

The concentration of neutral impurities in high-purity germanium far exceeds the concentration of the impurities P, Al, B and Cu so far discussed. This is, in part, a result of growing in hydrogen out of silica which has been chosen to reduce the electrically active impurity concentration. In crystals which have had no further thermal treatment after growth, the known neutral impurities appear in the following concentrations: $H \simeq 10^{15} cm^{-3}$ [20], $Si \simeq 10^{14} cm^{-3}$ [21], $O \simeq 10^{14} cm^{-3}$ [22] and $C \simeq 10^{13} cm^{-3}$ [23]. All of these elements are expected to remain electrically neutral in germanium. However, it has been discovered that complexes [24] between these elements and precipitates [25] can have electrical activity. The shallow centers which are formed by complexes of these elements and which are well understood are A(H,Si), A(H,C) and D(H,O) [17]. Further centers are the acceptors A_3, A_4, A_5, A_7 and several donor [17]. All of these centers are shallow (~ 10meV) and thus do not act as trapping centers at 77K but they may change $|N_A - N_D|$ to unacceptably high levels. Most of these centers are very weakly bound and dissociate or interconvert [26] at detector processing temperatures. The presence of these shallow complexes makes $|N_A - N_D|$ and with it the depletion voltage of detectors unpredictable. An originally excellent piece of ultra-pure germanium may turn out to be useless after inappropriate medium temperature processing such as n^+-contact formation with lithium diffusion. These complexes together with any copper can be removed by long time annealing in contact with liquid metal getters [27]. Such a process to be avoided, however, as gettering is costly and time consuming. Proper temperature time sequences during processing greatly reduce complex formation.

Another important center which leads to charge trapping is associated with "smooth pits" [25] as revealed by chemical etching. Whenever chemical etching shows a significant concentration (> $100 cm^{-2}$) of smooth pits, charge trapping is observed in detectors even in the absence of all the other well-known traps (Cu, V_2H, etc.). The smooth pits are always due to excess oxygen during growth due, for example, to air leaks or impure hydrogen. There are good intuitive and thermochemical arguments to support the suggestion that these pits have their origin in SiO_2 precipitates [25]. It is likely that the charge trapping is associated with the local lattice strain that accompanies the precipitation. DLTS analysis of crystals with smooth pits does not reveal any discrete levels. In this characteristic they are similar to crystals with a high density of dislocations. Both kinds of crystals have excessive local lattice strain, as revealed by chemical etching. In high-purity materials, chemical etchants do nothing more than decorate local variations in the chemical potential which can result from lattice strains.

EQUILIBRIUM THERMOCHEMISTRY

In a carefully reasoned and thorough study, Darken [28] applied the principles of equilibrium thermochemistry to the problem of stability of precipitated oxide phases in liquid germanium with special emphasis on SiO_2 precipitates and their interaction with aluminum. One of the central conclusions of this study was that under normal crystal growing conditions the reaction:

$$SiO_{2(1)} \rightleftarrows Si_{(1)} + O_{(1)} \tag{9}$$

is proceeding to the right because of the extremely low oxygen partial pressure. In another paper, Darken [29] showed that the oxygen in the melt will be in equilibrium with the oxygen in the vapor phase. These conclusions can be combined to show that if sufficient oxygen is added to the gas phase (leaks, impure H_2), reaction (9) will be driven to the left and SiO_2 precipitation can occur. This result is in accord with the experimental observation that SiO_2 precipitates (smooth pits) only occur in the presence of excess oxygen. Darken further shows that Al has a great chemical affinity for binding to SiO_2 but that the oxygen concentration in the melt will never be high enough for this reaction to proceed to form the aluminum silicate $Al_6Si_2O_{13}$ (mullite).

The behavior of SiO_2 in germanium as deduced from equilibrium thermochemistry still leaves unclear the process by which aluminum survives zone refining. It has been tacitly assumed that the aluminum survives zone refining by being absorbed on an external phase; namely, aluminum is associated with suspended particles of SiO_2 which do not agglomerate. This model was proposed because, among other reasons, it was thought highly unlikely that a dissolved chemical species would be non-segregating. Table I is an attempt to list some of the conflicting observations concerning Al survival. While equilibrium thermochemistry analysis has introduced some much needed limits to speculation by demonstrating which reactions cannot occur, there are still basic conflicts between proposed explanation and experimental observation.

TABLE I

Precipitate theory of aluminum survival	
Factors Favoring Theory	Factors Against Theory
1) Al does not seem to segregate in zone refining. 2) K=1 is improbable for any dissolved species. 3) Al has great affinity for SiO_2. 4) ^{14}C tracer studies [23] show that stable precipitates can survive generations of growth. 5) Al segregates normally if crystal is grown in C [32]. 6) Al segregates if excess Si is available [32].	1) Thermal chemistry shows that SiO_2 is a dissolving species during crystal growth. 2) No smooth pits observed if grown under dry conditions. 3) Zone refiner is "dryer" than crystal grower so SiO_2 dissolution should be more effective. 4) Al is not effectively removed by zone refining in C. 5) Reactions during crystal growth are near equilibrium as shown by equilibrium between O in gas and O in liquid.

Possible Scenarios

1) C is always contaminated with Al and precipitate theory is correct.
2) Al is already in Ge in particulate form but more stable than SiO_2, e.g., some precipitate of the form $Al_n(SiO_2)_m$.
3) There exists a dissolved chemical species of the form $Al_xGe_yO_z$ or $Al_xSi_yO_z$ which does not segregate and is stable in dry H_2 at 1200K.
4) SiO_2 precipitates do not dissolve during zone refining because of some peculiarity of kinetics during refining.

DISCUSSION

It turns out that the impurities Al, P and Cu are most often the cause for rejection of germanium crystals in the making nuclear radiation detectors. Cu does not form stable oxides and is easily removed from germanium. However, its fast diffusion in germanium and the possibility of the formation of volatile compounds requires careful attention to the purity of crystal grower parts, even those remote from the hot zone, and the purity of the H_2 gas. We found high Cu contamination in a closed-loop crystal grower used for tritium doping [20]. Insertion of a liquid nitrogen trap in the recirculating gas line solved the problem, leading to the conclusion that volatile Cu compounds were being transported by the hydrogen.

Phosphorus compounds are unstable and ubiquitous. We have found occasional P contamination of the crystal grower which is persistent until a general superficial cleaning. The P concentration in germanium crystals can often be correlated with the use of a particular silica crucible and it is suspected that the crucible is the principal source in such cases. Use of synthetic silica crucibles from different manufacturers do not lead to consistent results except for crucibles made by fusing natural quartz crystals (G.E.204) which always contain excessive P.

The chemistry of Al in germanium is very complicated and the mechanism by which Al survives zone refining is still in doubt. Equilibrium thermal chemistry analyses give valuable insights into how Al may react with Si-O-H in a Ge melt. However, its conclusions may be in conflict with observation. The theory may be correct if some peculiar kinetics of zone refining prevent SiO_2 precipitates from dissolving, e.g., if the RF-induced stirring rapidly transfers particles from the melting to the freezing interface with minimal contact with the melted zone.

By deduction, it is concluded that the main Al source is not the silica crucible. When the starting polycrystal does not contain electrically active Al (as measured by PTIS), the grown crystal will be essentially free of Al. The Al in the crystal appears to be proportional to the electrically active Al in the zone refined bar and is usually higher. SiO_2-Al complex formation is strongly supported by the observation that when Si is added to the melt, consequently pushing reaction (9) strongly to the left, Al segregates and has a higher concentration than in the zoned bar. One problem with this analysis is that the silica crucible itself is a strong getter for Al. This can be shown by attempting to grow Al-doped crystals in a silica crucible. A zone-leveled charge containing $10^{17}cm^{-3}$ electrically active Al produced a crystal with $2 \times 10^{12}cm^{-3}$ Al showing no segregation, i.e., of constant Al concentration.

When germanium is grown in carbon, Al segregates normally. It was not known initially whether the Al was coming from the carbon or from residual SiO_2-Al complexes in the germanium charge. In order to determine if Al was coming from the carbon crucible, two crystals were grown in carbon which showed normal Al segregation. The upper, seed end halves of these crystals were combined and a third crystal was grown. The resulting crystal showed normal Al segregation and precisely the Al concentration predicted from the now known Al concentration in the starting material. This means that for this particular crucible, there was no detectable Al contribution from the carbon.

The purity of the carbon and silica crucible materials is known only to the extent of general emission spectroscopy data on typical unfabricated materials from the manufacturers. The range of speculation about the mechanism of phosphorus and aluminum transport could be greatly limited by good analytical data on the actual or closely similar crucibles. Data from spark source or secondary ion mass spectroscopy could prove very useful.

Thermochemical analysis can be used to limit the number of reactions which can take place during crystal growing, but to be usefully applied, its limitations should be strictly observed. Aside from the problem of treating the kinetics of crystal growth as stated by Darken [28], there is an additional problem with the idealization of external phases. In the analysis of the system SiO_2-Ge-H_2, it was assumed that the external phases (SiO_2 and H_2) were in their pure states and would only be modified by reactions between them. But the SiO_2 is not a pure phase and contains about 1000ppm H_2O as measured by the OH content. The availability of this oxygen during crystal growth is completely unknown and may strongly modify the partial pressure of oxygen in the hydrogen ambient.

It is apparent now that most of the residual impurities present in ultrapure germanium crystals for detector applications arise from growing out of silica in hydrogen. These problems are the result of the added elements Si, O and H and their interactions with each other and with native defects and with any remaining electically active impurities. The selection of this crystal growing environment arose naturally as a solution to early purity problems. Crystals grown in vacuum were found to contain copper, so a high-purity gas was needed to shield the growing crystal. Crystals grown in carbon were found to have a high concentration of acceptor impurities, so carbon was replaced by synthetic silica. Crystals grown from silica in inert gas were found to contain precipitates (SiO_2) which caused carrier trapping, so the shield gas must be hydrogen. Dislocation-free crystals grown in hydrogen contained a deep trap, so crystals grown in hydrogen from silica must contain dislocations.

Although the above scenario results from a logical series of observations, subsequent knowledge suggests that this crystal growing method makes the production of the largest, highest purity crystals unreasonably difficult. Now that the dominant sources of impurities have been identified, re-examination of the original compromises made in selecting hydrogen-silica may make other methods attractive.

One of the first crystal grower environments tried for high-purity germanium was growing out of carbon in vacuum. This system eliminates all the problems related to H_2-silica (V_2H, SiO_2 precipitates, hydrogen complexes) and has additional benefits. However, the original reasons for abandoning carbon-vacuum become prominent, namely, copper from vacuum and shallow acceptors from carbon. It is now virtually certain that the acceptors come from Si-O-Al complexes in the germanium and not from the carbon. But even if the carbon is found to contain aluminum, it could likely be removed using in situ chlorine etching, as has been so successful when applied to cleaning silica in silicon MOS technology [30]. According to Darken's analysis [28], any SiO_2 precipitates should dissolve in refining under the same conditions--provided that carbon is not the source of the aluminum. In any event, the aluminum problem is now better defined so that it can be attacked directly.

Copper contamination while growing in vacuum must still be solved. Experience suggests that the proper choice of crystal grower materials may suppress this contamination. A great advantage of growing under vacuum is the elimination of the V_2H center. This means that dislocation-free crystals could probably be used for making detectors if no other deep vacancy level appeared. Once dislocation-free crystals are suitable, the thermal constraints during growth are greatly relaxed and large diameter crystals are more easily grown. The great resistance to the formation of the first dislocation is shown by the routine production of large diameter, dislocation-free silicon crystals, even under the very poor thermal conditions of floating zone growth.

ACKNOWLEDGEMENTS

This work has benefited greatly from the continuous interest and support of F. S. Goulding.

This work was supported by the Director's Office of Energy Research, Office of Health and Environmental Research, Pollutant Characterization and Safety Research Division of the U.S. Department of Energy under Contract No. DE-AC03-76SF00098.

REFERENCES

1. E. E. Haller and F. S. Goulding in: Handbook on Semiconductors, C. Hilsum ed. (North-Holland 1980) Vol. 4, Ch. 6C.

2. E. M. Pell, J. Appl. Phys. 31, 291 (1960).

3. R. N. Hall in: Semiconductor Materials for γ-Ray Detectors--Proceedings of the Meeting, W. L. Brown (BTL) and S. Wagner (BNL) eds. (1966) p. 27.

4. G. S. Hubbard, E. E. Haller and W. L. Hansen, IEEE Trans. Nucl. Sci. NS-25, No. 1, 362 (1978).

5. R. N. Hall and T. J. Soltys, IEEE Trans. Nucl. Sci. NS-18, No. 1, 160 (1971).

6. W. L. Hansen, Nucl. Instr. and Methods 94, 377 (1971).

7. J. Czochralski, Z. Phys. Chem. 92 219 (1918).

8. C. Kittel, Introduction to Solid State Physics (John Wiley 1968) 3rd ed. p. 561.

9. A. G. Tweet, J. Appl. Phys. 30, 2002 (1959).

10. H. Föll and B. O. Kolbesen, J. Appl. Phys. 8, 319 (1975).

11. R. N. Hall and T. J. Soltys, IEEE Trans. Nucl. Sci. NS-18, No. 1, 160 (1971).

12. E. E. Haller, G. S. Hubbard, W. L. Hansen and A. Seeger, Inst. Phys. Conf. Ser. No. 31, 309 (1977).

13. P. Glasow and E. E. Haller, IEEE Trans. Nucl. Sci. NS-23, No. 1, 92 (1976).

14. G. Ziegler, Z. Naturforsch. 169, 219 (1961).

15. T. F. Ciszek in: Semiconductor Silicon, E. L. Kern and R. R. Haberecht eds. (The Electrochemical Soc. 1969) p. 156.

16. G. S. Hubbard, E. E. Haller and W. L. Hansen, IEEE Trans. Nucl. Sci. NS-26, No. 1, 303 (1979).

17. E. E. Haller, W. L. Hansen and F. S. Goulding, Adv. in Physics 30, No. 1, 93 (1981) and references therein.

18. W. L. Hansen and E. E. Haller, IEEE Trans. Nucl. Sci. NS-19, No. 1, 260 (1972).

19. W. D. Edwards, J. Appl. Phys. 39 1784 (1968); ibid. 39 2457 (1963).

20. W. L. Hansen, E. E. Haller and P. N. Luke, IEEE Trans. Nucl. Sci. NS-29, No. 1, 738 (1982).

21. A. J. Tavendale, Australian Atomic Energy Comm., private communication.

22. R. J. Fox, IEEE Trans. Nucl. Sci. NS-13, No. 3, 367 (1966).

23. E. E. Haller, W. L. Hansen, P. N. Luke, R. McMurray and B. Jarrett, IEEE Trans. Nucl. Sci. NS-29, No. 1, 745 (1982).

24. E. E. Haller, B. Joós and L. M. Falicov, Phys. Rev. B 21, 4729 (1980).

25. R. N. Hall, IEEE Trans. Nucl. Sci. NS-19, No. 3, 266 (1972).

26. E. E. Haller, Inst. Phys. Conf. Ser. No. 46, 205 (1979).

27. R. N. Hall, IEEE Trans. Nucl. Sci. NS-21, No. 1, 260 (1974).

28. L. S. Darken, IEEE Trans. Nucl. Sci. NS-26, No. 1, 324 (1979).

29. L. S. Darken, J. Electrochem. Soc. 126, 827 (1979).

30. E. H. Nicollian and J. R. Brews, MOS Physics and Technology (John Wiley 1982) p. 764.

31. J. H. Crawford and L. M. Slifkin, Point Defects in Solids, Vol. 1 (Plenum Press 1972).

32. E. E. Haller, W. L. Hansen, G. S. Hubbard and F. S. Goulding, IEEE Trans. Nucl. Sci. NS-23, No. 1, 81 (1976).

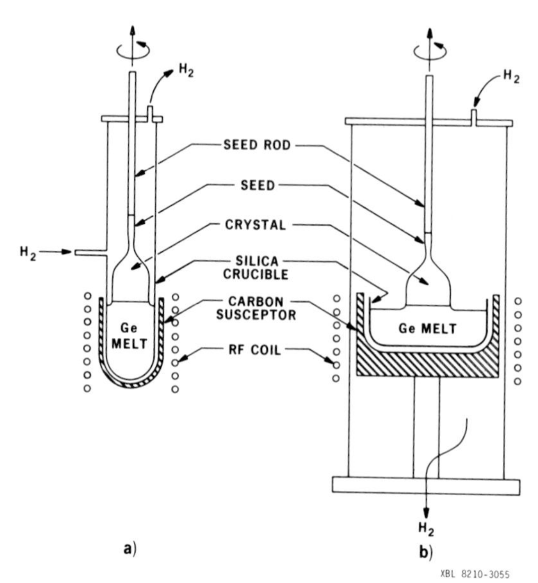

Fig. 1. Simplified illustrations of crystal grower designs currently used for high-purity germanium. Design (a) [5] uses an external carbon susceptor and a crucible which closely matches the crystal diameter. The shape of the melt-solid interface is probably dominated by heat radiation from the crystal. Design (b) [6] uses an internal susceptor and a large melt diameter. The shape of the melt-solid interface for this design is probably dominated by heat transport by hydrogen gas convection.

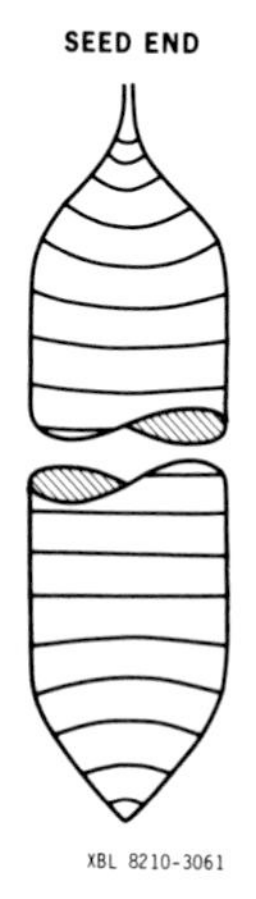

Fig. 2. The desired evolution of the melt-solid interface during crystal growth to minimize thermal stress so as to achieve a uniform dislocation concentration. This is a necessary but not sufficient condition for low thermal stress because the crystal can still experience a large radial thermal gradient after it is grown but while it is still hot.

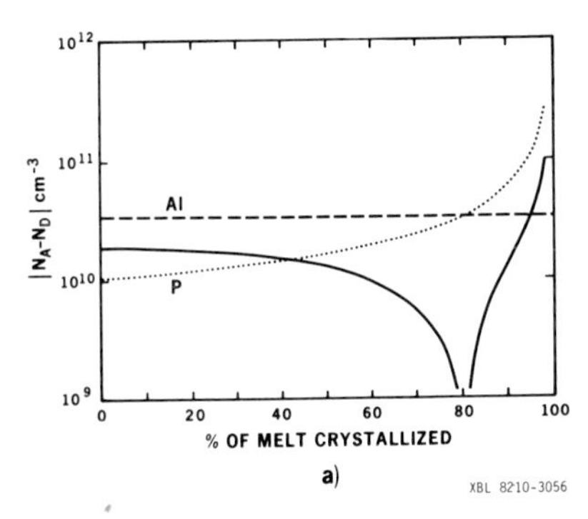

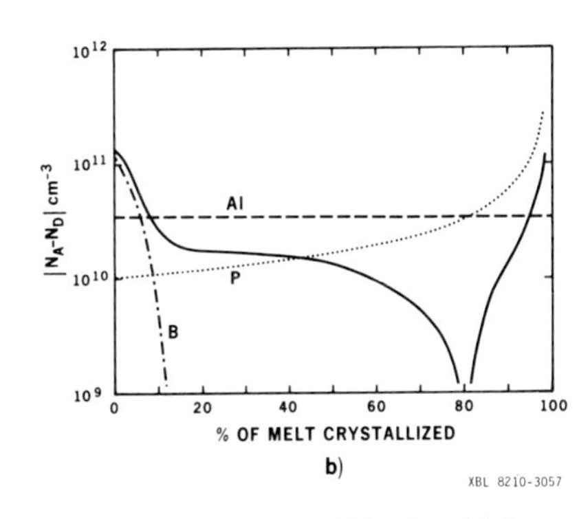

Fig. 3. (a) The solid line shows a common type of impurity profile for high-purity germanium. This profile results from equal concentrations of Al and P in the melt of 3 x 10^{10}cm^{-3}. The Al is non-segregating and the P has an effective segregation coefficient of 0.3 at our growth rate. (b) This profile is a result of the same P and Al impurity concentrations as 3a with the addition of some boron. Boron disappears faster than its segregation coefficient would predict due to oxidation by the silica crucible.

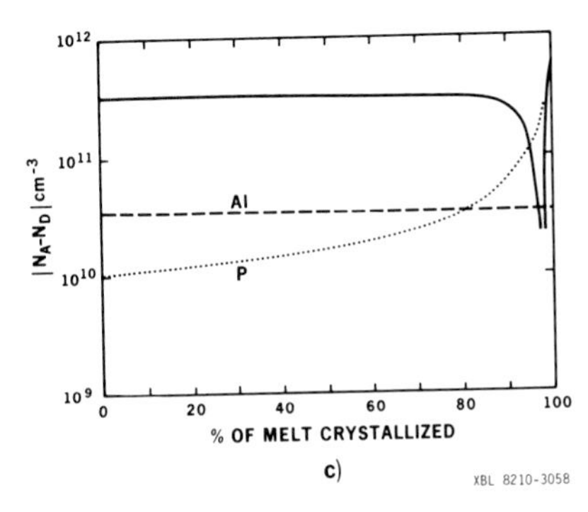

Fig. 3(c) A crystal with the same impurities as 3a but dislocation free. The constant acceptor concentration of 3 x 10^{11}cm^{-3} is due to (V_2H) complex formation.

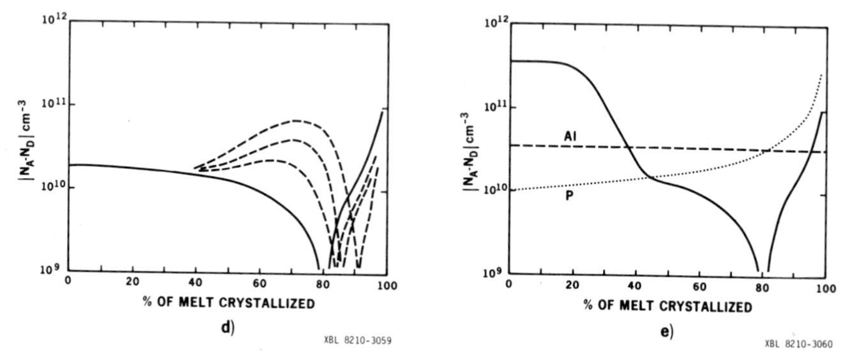

Fig. 3(d) Profiles of crystals with the same impurities as 3a but which have "coring". Coring is a radial impurity gradient which distorts the resistivity measurement. This radial impurity gradient is always more strongly p-type toward the outer surface and less p-type or n-type toward the center. (e) A more complicated profile which is the result of the same impurities as 3a but which is dislocation free for the first 20% of growth. Dislocations which appear in the last 80% act as sinks for excess vacancies which prevents (V_2H) formation.

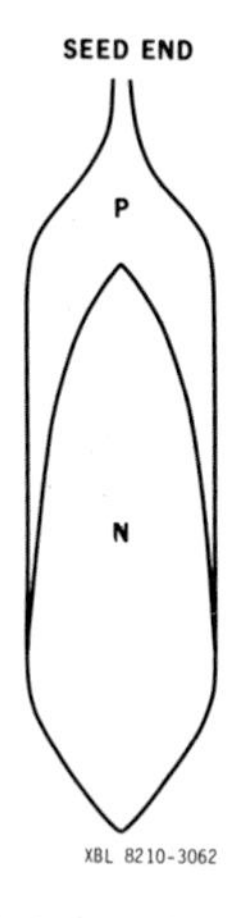

Fig. 4. The apparent impurity distribution inside a crystal which shows "coring" and which would be otherwise high-purity n-type. The coring phenomenon is often the result of copper contamination of the growing crystal through the gas phase. The shape of the acceptor distribution is due to the time-temperature history of the copper diffusion.

SILICON MATERIAL GROWTH FOR NUCLEAR RADIATION DETECTORS

P. A. GLASOW AND B. O. KOLBESEN
Central Research Laboratories, Siemens, AG, Erlangen and München,
F. R. Germany

INTRODUCTION

As a base material for semiconductor devices, silicon is more widely used than any other semiconductor. The physical properties, in particular the bandgap which is significantly larger than that of germanium, makes the material extremely important for electronic devices. The world's total annual production of silicon is at present some 2000 t [1]. Compared with this, the 10 kg/year of silicon that is used for detectors is rather modest. However, since work on semiconductor radiation detectors started 25 years ago, silicon in addition to germanium forms the centre of interest as the basis for production of nuclear radiation spectrometers, mainly as high energy particle detectors, but also as X-ray detectors. Today, silicon detectors are well established in experimental physics and are widely used in technical applications [2]. As Figure 1 shows, detector grade silicon up to 3" diameter can now be made which has outstanding characteristics. Highly developed particle detectors with up to 35 cm^2 sensitive area and 5 mm thickness, manufactured in our laboratory, exhibit better than 35 keV FWHM α-resolution at room temperature and below 12 keV for high energy betas at liquid nitrogen temperature, whereas small, 80 mm^2 area lithium-drifted silicon detectors (Si(Li)) with 140 eV for ^{55}Fe x-rays are widely used in energy dispersive x-ray analyzers.

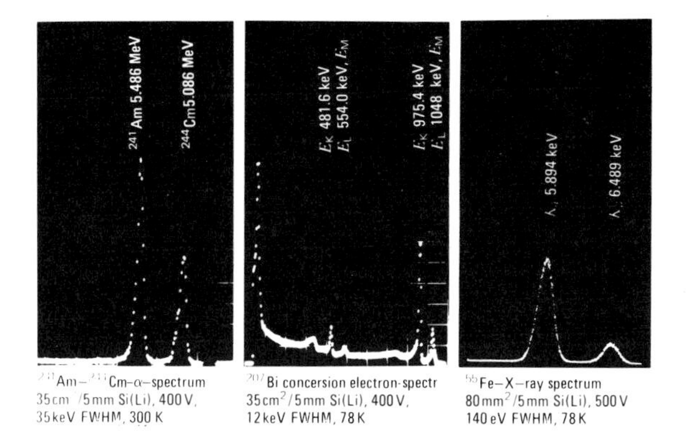

Fig. 1. Spectra obtained with lithium-drifted silicon detectors.

One of the primary factors in the performance of the detector is the quality of the starting material. In contrast to germanium, where the breakthrough in the technology as a detector grade material came with the development of the high purity germanium specifically for detectors at the General Electric and the Lawrence Berkeley Laboratories [3, 4], silicon detectors derive from the material development carried out industrially for other electronic devices. Silicon is seldom directly produced as a bulk material for detectors; it

Mat. Res. Soc. Symp. Proc. Vol. 16 (1983)

is obtained almost accidentally. Therefore it is sometimes a problem to obtain high quality silicon permitting the production of detectors with high breakdown voltage and low charge trapping, especially since, except with the specific resistance, it is still uncertain in the end which material parameters are really necessary for good detector performance.

Within the scope of this symposium, we should like to give a survey of the manufacturing steps of high purity silicon, as it is used for radiation detectors, silicon with specific resistivities from 1000 to 80,000 Ohmcm, and then discuss the crystal properties, mainly the inhomogeneities, which can have an influence on the detector behaviour.

SILICON PURIFICATION AND INGOT GROWTH

Reduction of SiO_2 Metallurgical Silicon Purification

Figure 2 shows a flow diagram for the production of semiconductor grade silicon. For an indepth study, reference is drawn to the books by Hadamovski 1972 [5], Pfann [6], and especially for floating zone silicon to the excellent text by Keller and Mühlbauer [7], from which some of the illustrations here are taken. The starting material is quartzite gravel or crushed quartz. The quartz is reduced with coke to yield metallurgical grade silicon of moderate (98%) purity. The metallurgical silicon is converted into the intermediate chemical compound $SiCl_4$ and particularly $SiHCL_3$ and further purified by

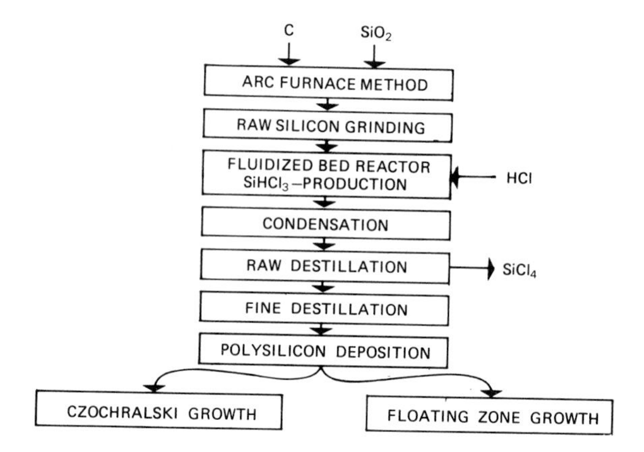

Fig. 2. Silicon material technology.

multiple fractional distillation. The metal chlorides can thereby be relatively easily separated, while the vapour pressures of the critical impurities PCl_3 and BCl_3 lie very close to those of $SiHCl_3$. With further refining of the silicon using the floating zone refining method, the presence of boron with its segregation coefficient, $k = 0.8$, becomes problematical since almost no further purification of boron is possible. For high purity silicon, $BHCl_3$ ought to be therefore extensively eliminated prior to zone refining; P with $k = 0.35$ can be largely eliminated by repeated zone passes. The final semiconductor grade polycrystalline silicon is obtined by chemical vapour deposition on the thin hot silicon or graphite cores in the so-called Siemens reactor by reduction of the purified $SiHCl_3$ in H_2. The deposition rates are about 1 kg/h today. This silicon has already high purity, commonly it has n-type conductivity due to phosphorus impurities. Low concentrations of electrically active impurities of the order of 1×10^{11} cm^{-3} corresponding to up about 50,000 Ωcm resistance can be obtained.

Another purification technique has been discussed in the last few years by Yusa et al [8]. They use monosilane SiH_4, which is very inflammable, as a source gas instead of $SiHCl_3$ and $SiCl_4$, and is purified by means of a special molecular sieve method. The advantage of monosilane for the production of high purity silicon is that monosilane decomposes at about 850° C and thus at a lower temperature than $SiHCl_3$ (1150° C). Monosilane is also much less corrosive. Finally, in the process for producing monosilane, one of the by-products, boron hydride, is extensively removed by an associated chemical reaction The residual boron content in silicon crystals prepared from SiH_4, is claimed to be as low as 0.02 - 0.01 ppba (atomic ratio), i.e., 5×10^{11} acceptors cm^{-3} or 25,000 Ωcm p-Si [8].

Single Crystal Growth

Polysilicon is convereted to a single crystal by two methods: the Czochralski method and the floating zone technique.

Czochralski Growth Technique. Figure 3a shows the principle of the Czochralski crystal growth technique. The polysilicon is molten in a quartz crucible which is placed in a carbon susceptor for stability reasons. A seed crystal is lowered into the melt and slowly withdrawn, while being rotated. Single crystals up to 6 inches in diameter and more than 1 meter in length can be grown today. The market share of Czochralski grown crystals is about 80 - 90%. However, because of its high melting point at 1415° C and the high oxidation potential of silicon, the silicon melt is highly corrosive. Quartz crucibles soften at 1415° C and, in addition, the molten silicon extracts large quantities of oxygen and other impurities from the quartz. Quartz crucibles are currently used to obtain Czochralski silicon crystals, however, the oxygen incorporation into the hot silicon is tolerated because of the overall cost efficiency of the Czochralski process.

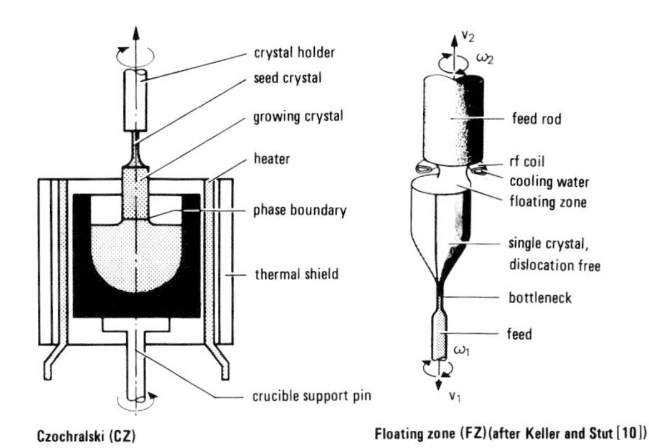

Fig. 3. Growth methods for silicon single crystals (schematic).

Floating Zone Growth Technique. High purity silicon single crystals with resistivities up to 50,000 Ωcm (the world record lies at 100,000 Ωcm) are made by the floating zone process (Figure 3b). With floating zone growth, a localized molten region is produced by high frequency coils around the vertically oriented crystal and drawn along the poly-crystalline rod either by movement of the rod or the heat source. Thus, the floating zone has no contact with foreign matter; it is held in place between the melting and growing interface by surface tension, interface tension and electrodynamic forces of the rf coil. The growth environment is vacuum or an argon atmosphere. A big advantage of the floating zone method is that it can be repeated several times in the same direction, so that the impurities are drawn to one crystal end. Finally a single crystal is obtained with a crystallographic orientation, which is given by the single crystal seed orientation. Floating zone silicon has a market share between 10 - 20%.

Two fundamental modifications to the FZ method are shown in Figure 4 that produce floating zone single crystals. The bottom- and top-seeded processes differ mainly in growth direction. To achieve cylindrical growth, it is generally necessary to have a slightly positive melt meniscus angle. For bottom seeded growth, this positive angle is easily obtained because of the gravity overhang of the melt at the lower rim of the melting zone (Figure 4a). For top seeded growth, however, the overhang is obtained only at the melting interface, whereas the growing interface tends to neck in. Only by keeping the melting zones extremely short is it possible to maintain stability here. For bottom seeded growth the situation is much more favourable. The part which is just melting near the crystal is necked in. This tapering allows the use of a coil with a diameter smaller than that of the rod itself. This technique is called the needle-eye method and is applied successfully for the growth of larger diameter crystals. Thus, it is possible to produce single crystals with 10 cm diameter using a 3.4 cm diameter coil.

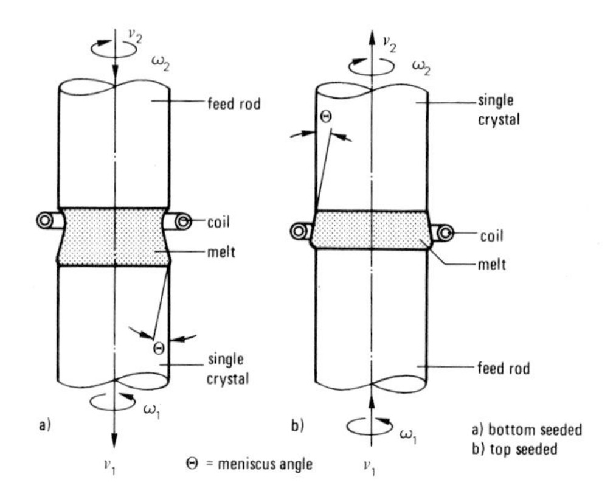

Fig. 4. Floating zone methods.

To get dislocation free growth, the thin-seed bottle-neck technique [11,12] is now in general use, in which a seed with a small diameter is used and a bottleneck is made at the fusion point of the seed crystal and the feed rod by means of rapid pulling (Figure 5). The dislocations disappear and do not return even after the crystal has grown to its full diameter.

However, the crystal weight upon a thin seed with an even thinner bottleneck, is limited to about 4 kg. To improve the bottle-neck growth stability, a support system for the crystal cone must be introduced. Figure 6 shows the principle of such a system [13] (other systems are also possible [14]). After growing a certain length of the dislocation-free crystal, a funnel that rotates together with the seed is slid upward along the rotating shaft until it surrounds the seed and cone region of the crystal. Then the funnel is filled with small balls. They adapt nicely to any shape regardless of whether or not the cross section is circular [7].

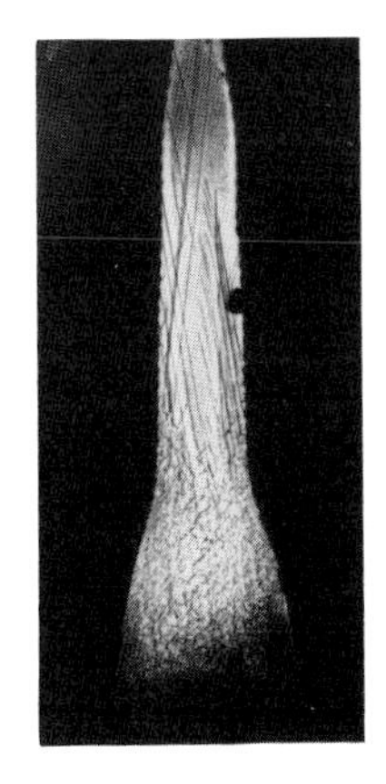

Fig. 5. Dislocation-free growth by bottle-neck technique.

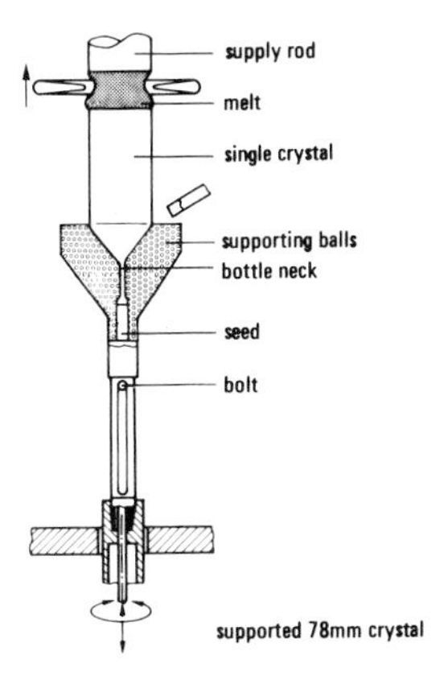

Fig. 6.
Crystal supporting system [7].

The Czochralski and float zone methods are compared [15] in Table I. By the Czochralski technique, large ingots with diameters of 6" and more and with large weights of 60 kg may be produced. However, large diameter ingots are also possible today, using the floating-zone technique. The mechanical growing conditions of the Czochralski method are favourable since, the crystal hanging on the thin seed is stable, whereas the floating zone crystal standing on the seed has to be supported. The purity of the floating zone material is higher, and therefore floating zone silicon is used for power devices, for devices with high operating voltage, for thyristors, and it is also the starting material for nuclear radiation detectors. Czochralski grown crystals are used for low voltage devices, e.g. ICs, telecommunication, entertainment components, and last but not least for solar cells. The overall material costs are about the same for both processes.

TABLE I
Czochralski and floating-zone silicon
Comparison of growth methods after Keller and Mühlbauer (1981) [15]

	CZ	FZ
purity	poor (oxygen) one pass	high several passes possible
crystal size diameter weight	large easy 150 mm 20-60 kg	large possible 100-125 mm 20-40 kg
mech. growth conditions	gravity stabilizes melt and melt and crystal	gravity destabilizes melt and crystal
thermal growth conditions	small gradients	large gradients
economy	large weights low pull rate < 2mm/min if resistance heated	reasonable weights fast pull rate > 3mm/min RF - heating necessary no crucible
	overall costs	about the same
application	low voltage devices	high voltage devices power devices nuclear radiation detectors
market share - total	80%	20%
IC's	> 95%	< 5%

In Table II, important features of Czochralski and floating-zone silicon are compiled. Silicon with high resistivity and minority carrier lifetime is needed for nuclear radiation detectors, and is obtained only by the floating-zone method. The upper limit of the specific resistivity obtainable by the floating-zone method is determined by the polysilicon production method. The influence of O and C on detector properties is not yet well understood.

Precise doping of crystals

In order to obtain surface barrier detectors with large depletion depth, silicon crystals with high resistivity up to 30,000 Ωcm (n-type, sometimes p-type) are required. The resistivity of the starting material commonly used for lithium-drifted detectors is about 1000 Ωcm p-type. In addition to having a low net-to-absolute impurity concentration, the residual dopants should be homogeneously distributed in the single crystal. The incorporation of the dopants from the liquid zone into the growing crystal is controlled primarily by the effective segregtion coefficient of the dopant being used. However, despite similar segregation coefficients, there are marked homogeneity differences between group III and group V dopants in as-grown crystals; dopants from the group III lead, in general, to smaller dopant variations in the growing crystal than do group V impurities.

TABLE II
Czochralski and floating-zone silicon
Comparison of important properties

	CZ	FZ
resistivity range p–type, n–type	0.005–50 Ωcm	0.1 > 50000Ωcm
dopants	B, P, As, Sb	B,P
life time	10–50 μs	100–5000 μs
defect densities	high	low
capability for internal gettering	yes	no
oxygen range	$2\times10^{17}-2\times10^{18}$ cm^{-3}	$<5\times10^{14}-2\times10^{16}$ cm^{-3}
concentration average	8×10^{17} cm^{-3}	5×10^{15} cm^{-3}
carbon range	$<5\times10^{15}-5\times10^{17}$ cm^{-3}	$<5\times10^{15}-3\times10^{17}$ cm^{-3}
concentration average	4×10^{16} cm^{-3}	2×10^{16} cm^{-3}

The equilibrium segregation coefficient k, derived from the binary-phase diagram are < 1 for all important dopants in silicon and are valid only for growth rates so small that total homogeneity of the dopant in the melt, (either by diffusion or convection) can be assumed. This fact is illustrated in Figure 7. The left-hand side depicts the equilibrium case with a negligible growth rate. The right-hand side shows a non-equilibrium but steady-site case with a constant finite growth rate. In the melt near the interface a liquid region with enhanced impurity concentration is formed which is called diffusion layer. The impurity atoms that are not accepted by the solid phase have not enough time to diffuse entirely back into the bulk of the liquid, thus creating an enhanced concentration in this region. The effective segregation coefficient and therefore the doping concentration is larger, the thicker the diffusion layer.

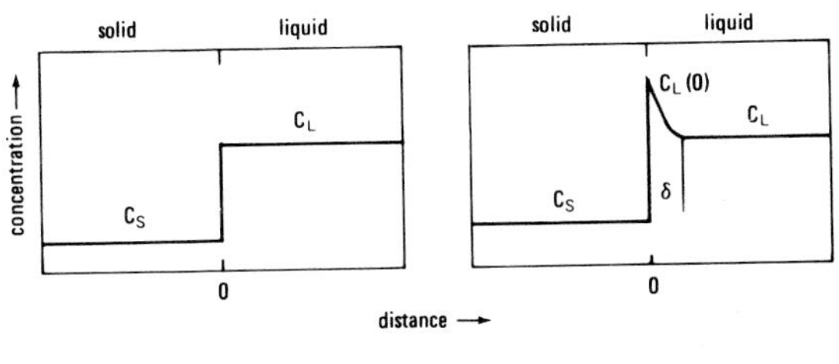

Fig. 7. Impurity concentration near the growing interface.

Figure 8 shows a model of a floating zone of a 53 mm diameter crystal for growth conditions without facets (i.e., <100> seed orientation). Above a certain seed rotation rate a flow torus forms in the molten zone. This reduces the diffusion layer thickness in a ring between the rim and the center of the growing interface in the zone of increased melt flow, while a stagnation circle builds up in the center and a stagnation ring at the interface periphery. Because the dopant concentration varies with the thickness of the diffusion layer, thick in the center and at the rim (Figure 7b), the resistivity has two maxima between the periphery and the crystal center. This effect is more pronounced with larger crystal diameters.

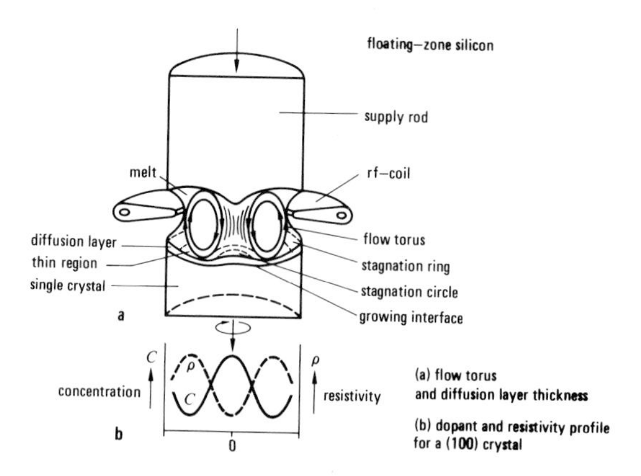

Fig. 8. Model of a flow pattern and dopant distribution at floating zone [7].

The variation of the dopant concentration C_S in the solid resulting from changes in diffusion layer thickness is:

$$\frac{\Delta C_S}{C_S} = \frac{1}{k} \cdot \frac{dk}{d\delta} \cdot \Delta\delta = \frac{d(\ln k)}{d\delta} \cdot \Delta\delta \tag{1}$$

Here, the relative change of dopant concentration is proportional to the slope of the semi-logarithmic curves k as a function of δ (Figure 9 [7]). For technical growth conditions, the diffusion layer thickness lies in the range of a few hundred micrometers. The curves in Figure 9 indicate that radial dopant variations of group V dopants must increase as the k_O of the dopant decreases. For group III impurities, however, in spite of their considerably different equilibrium distribution coefficients, rather similar radial dopant profiles have to be expected, because their effective distribution coefficient varies but little with the diffusion layer thickness δ. Additionally, the vapour pressure of the residual impurity phosphorus atoms evaporate out of the melt surface and this adds additional nonuniformity of the impurity distribution. A zone refined p-type silicon crystal, however, produced by repeated zone refining, can have an excellent uniformity in resistivity distribution if it is not compensated. Therefore, suitable p-type silicon with uniform boron distribution, for lithium-drifted detectors could be produced easily.

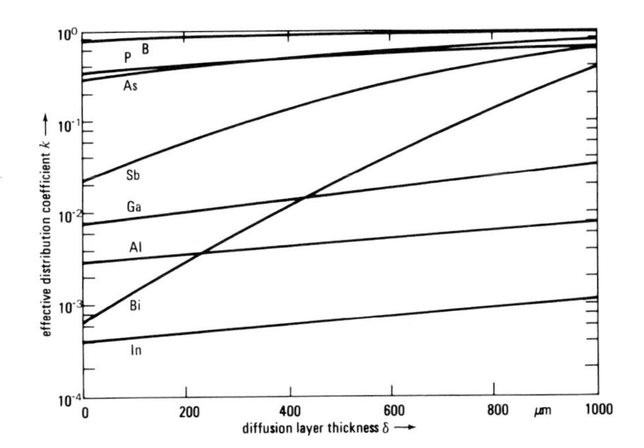

Fig. 9. Calculated relationship between diffusion layer thickness and effective distribution coefficient for several dopants.

Neutron transmutation doping

High resistivity n-type silicon is produced from p-type material by neutron transmutation doping. Only by this low temperture method can one obtain both macroscopic and microscopic homogeneity and phosphorus doping concentrations reproducible to better than ± 5%. This homogeneity is required not only for detectors, but also for high power devices, rectifiers and thyristors.

Neutron transmutation doping was described in principle in 1951 [16]. However, it was thought for a long time that a practical realization was impossible due to the radiation damage of the lattice. It was only in 1973 that Schöller and Haas [17] at Siemens, and Herzer at Wacker took up the process again and brought it to maturity for device application. Unfortunately, the process can only be used to introduce one dopant, phosphorus.

The principle of neutron transmutation doping is that phosphorus donors can be created in silicon very homogeneously by the nuclear reaction in which thermal neutrons are captured by ^{30}Si. The transmutation of this isotope, which is 3% abundant in Si, into P occurs in two steps:

$$^{30}Si + \text{neutron} \dashrightarrow {}^{31}Si \dashrightarrow {}^{31}Si + \gamma \dashrightarrow {}^{31}P + e^- + \bar{\nu}_e. \qquad (2)$$

With the capture of a neutron, the ^{30}Si nucleus is transformed into an excited ^{31}Si nucleus which decays to its ground state by emission of a γ quantum. The ^{31}Si nucleus in the ground state is not stable, but beta decays into the stable ^{31}P nucleus. The half life of this beta decay is 2.62 h.

There is a possibility for some slight radioactivity caused by the reaction [18] during irradiation:

$$^{31}P(n,\gamma) \dashrightarrow {}^{32}P \xrightarrow{14\ \text{days}} {}^{32}S + e^- + \bar{\nu}_e. \qquad (3)$$

where the relatively short decay time of ^{32}P seems to have an almost negligible effect on the performance of a detector, except for detectors for low background applications.

Figure 10 shows the excess donor/acceptor concentration as a function of irradiated neutron dose for high purity, detector grade material [19]. The phosphorus concentration Cp obtained depends upon the irradiation time t and the thermal neutron flux density Ø according to

$$Cp = K^x \, \emptyset \, t \qquad (4)$$

where K^x represents an effective cross-section which lies in the range 1.5 x 10^{-4} to 2.2 x 10^{-4} atom/neutron cm, depending on the reactor used. For example, in a reactor with a flux density of 2 x 10^{13} neutrons/cm^2 sec the irradiation time required to obtain a resistivity of 170 Ωcm (Cp = 2.6 x 10^{13} cm^3) is about 2 h.

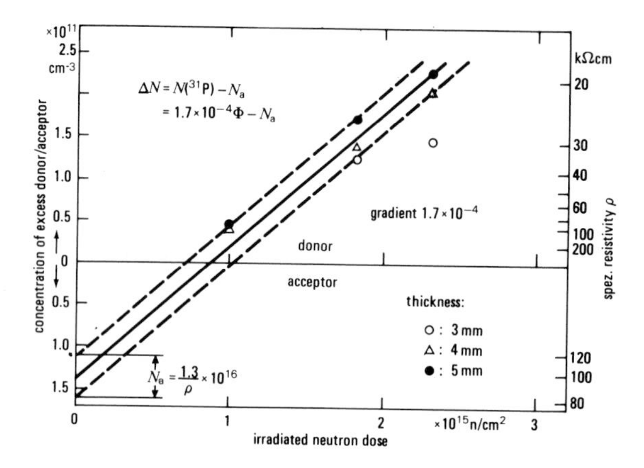

Fig. 10. Excess donor/acceptor concentration as a function of irradiated neutron dose [17].

Figure 11 [19] shows that for a high doping accuracy it is absolutely necessary to have a precisely controlled quantity of neutrons and, therefore, a well-adjusted neutron flux density in the irradiation position. Several nuclear reactors in the FRG are suitable for bombarding silicon samples up to 100 mm in diameter and 500 mm in length, where a thermal neutron flux density of a homogeneity better than ± 5% is guaranteed.

Crystal perfection is degraded by the irradiation mainly by the fast neutrons contained in the thermal flux. In order to bring the phosphorus atoms generated by the nuclear reaction into lattice sites and to restore the lattice perfection, the samples have to be carefully annealed at temperatures of about 700 °C after neutron transmutation doping.

Other doping techniques, such as pill doping or gas doping, do not play any role in the production of detector grade silicon.

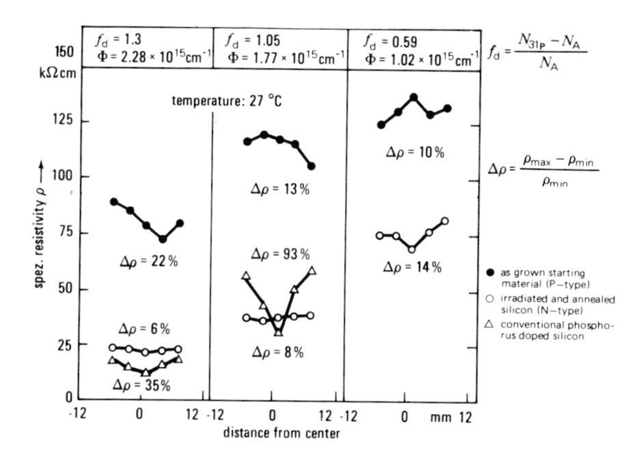

Fig. 11. Profiles of resistivity distributions [17].

INHOMOGENEITIES AND CRYSTAL DEFECTS, ORIGIN AND CHARACTERIZATION

As with today's integrated circuits and power devices, semiconductor nuclear radiation detectors also demand single crystals of high perfection. Defects and inhomogeneities of impurities can deteriorate the electrical detector characteristics and consequently the yield of a device. At present, silicon crystals grown by the Czochralski (CZ) or floating zone (FZ) method are grown dislocation-free. The main defects and inhomogeneities occurring in such crystals are striations, swirl defects, and precipitates of impurities (for an in-depth study see [20,21], containing comprehensive literature surveys).

Impurities and Intrinsic Atomic Point Defects

Apart from dopants which are intentionally incorporated into silicon crystals, oxygen and carbon are the main residual impurities in electronic-grade silicon. Figure 12 presents the typical ranges of oxygen and carbon concentrations in CZ and FZ crystals taken from various commercial suppliers determined by infrared spectroscopy. The oxygen and carbon concentrations are up to 10^{16} cm^{-3}, which is high compared to normal doping concentrations of 10^{11} or 10^{12} cm^{-3}. Most striking in Figure 12 is the considerable difference in the oxygen content of CZ and FZ crystals. In CZ crystal growth the quartz crucible is attacked by the silicon melt resulting in the dissolution of large amounts of oxygen in the melt [23]. A certain fraction of this oxygen is incorporated into the growing crystal on interstitital lattice sites.

The carbon in CZ and FZ crystals originates mainly from the polycrystalline starting silicon [22]. In the past, a considerable fraction of the carbon in CZ crystal growth resulted from graphite parts of the crucible heating system [24]. At that time, the average carbon concentration in CZ crystals usually exceeded 10^{17} cm^{-3}.

A few years ago, we tried to find out whether dopants, compensation degree or neutral impurities affect the detector performance and the Li-drift

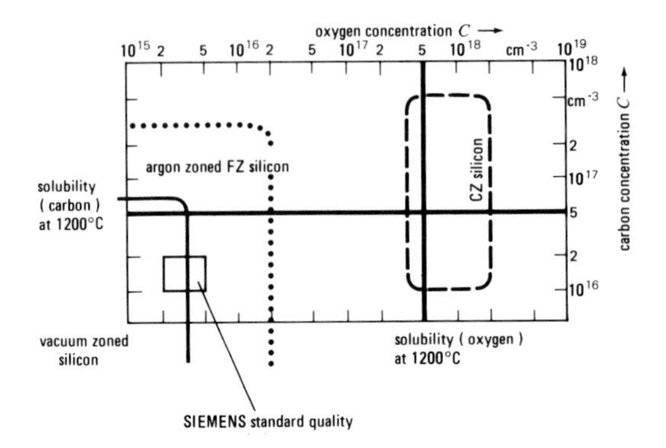

Fig. 12. Interstitial oxygen and substitutional carbon concentrations of CZ and FZ silicon crystals of various suppliers determined by infrared spectroscopy according to DIN (reproduced from ref. 20).

velocity. For this purpose, samples were taken from various p-type silicon crystals, which were purchased from various suppliers and from which the small 12 mm diameter, 3 mm thick lithium-drifted detectors were fabricated for energy dispersive x-ray analysis. From those samples the [P], [B], |[P] - [B]| and [O] concentrations were determined by infrared spectroscopy at low temperatures, as described in [25]. Finally the impurity concentration has been compared with the energy-resolution for ^{55}Fe X-rays. Table III summarizes the results: The detectors are evaluated according to their energy resolution, whereby:

+ means: a large number of the detectors have FWHM $\leq$ 130 eV.
- means: poor resolution; hardly any detector achieves less than 140 eV FWHM.

The detector resolution has been measured by pulsed optical feedback preamplifier system [26].

The detectors show clearly differences correlated with the crystals. Comparison with the IR measurements show however, that the [P], [B] and [O] concentrations or the compensation are not correlated to the detector performance and the lithium-drift-behaviour. The material giving the worst results had, in fact, the lowest [O] concentration, while crystal No. 54203 with a high [O] content produced good results.

In similar tests it was not possible to establish any connection between the residual carbon content and detector properties. However, it is not known, which defects can be created by the presences of oxygen and carbon in silicon, which would influence the detector properties.

Apart from impurities, a certain concentration of intrinsic atomic point defects (vacancies and self-interstitials) exist in a crystal lattice [28]. At present, the controversy is still going on as to whether vacancies or silicon self-interstitials are the dominant atomic point defects in silicon at high temperatures [29], even though experimental evidence clearly supports the

TABLE III
Impurity concentration of different silicon crystals, measured by low temperature infrared spectroscopy [23] and classification of the spectrometric properties as X-ray detectors (A1-B1) and particle detectors (A5-C1).

Crystal	Diameter mm	O-conc. N_O $\times 10^{15}$ cm^{-3}	C-conc. N_C $\times 10^{15}$ cm^{-3}	B-conc. N_B $\times 10^{12}$ cm^{-3}	P-conc. N_P $\times 10^{12}$ cm^{-3}	$\lvert N_B - N_P \rvert$ $\times 10^{12}$ cm^{-3}	ρ Ωcm	class.	T K
A 1	38	< 1	–	20	4.2	16	850	—	78
A 2	28	15	–	14	7.0	9	1500	•	78
A 3 seed	41	6	–	14	3.0	13	1250	•	78
A 3 tail	–	6	–	13.5	2.3	9	1500	•	78
A 4 seed	36	19	–	17.7	8.4	9	1500	+	78
A 4 tail	–	20	–	18	8.5	9.5	1400	+	78
B 1	36	2.9	–	17	2.3	15	950	+	78
A 5	75	1.2	4.7	13	4.6	8.7	1100	•	300
A 6	75	0.16	< 5	9	3.4	6.6	1000	•	300
B 2	75	< 0.1	6.5	8.9	3.7	5.2	2500	+	300
B 3	75	0.2	6.4	11	2.6	7.5	1800	+	300
C 1	52	1.9	11	63	3.4	29	450	•	300

dominance of self-interstitials. The estimated concentration of self-interstitials evaluated from a number of experiments is about 2 x 10^{16} cm^{-3} [30]; the concentration of vacancies may be of the same order of magnitude or less [20]. Their influence on the detector properties is not yet known.

Defects and Inhomogeneities

At present CZ and FZ single crystals, in particular of large diameter (> 2"), can be grown free of line dislocations and macroscopic defects such as twins and stacking faults. The influence of dislocations on the minority carrier lifetime in silicon surface barrier detctors has been discussed in [31]. The main residual defects and inhomogeneities which may occur are:

1. Striations
2. Swirl defects

Striations

Striations are microscopic inhomogeneities of the impurity distribution in silicon crystals, periodically fluctuating axially and/or laterally. Striations of electrically active impurities, e.g. phosphorus, can easily be made visible by approprite etching [32], using spreading resistance measurements or the SEM-EBIC (scanning electron microscope-electron beam induced current) technique [33]. Striations of electrically inactive impurities can be revealed by x-ray topography methods, if their concentrations are sufficiently high (> 10^{16} cm^{-3}).

We have found striations in silicon after lithium drift. Figure 13 shows striations, revealed by copper staining in the mesa of a small lithium drifted diode. Also shown are the axial variations of the specific resistivity, measured by the spreading resistance technique along three lines in 25 µm steps. The specific resistivity varies across the striations by a factor of more than 100 between > 10^5 and 10^3 Ωcm, which is the resistivity of the starting material. The striations are periodically separated by the interval λ = 250 µm. The interval λ is equal to v/w, where v is the pulling rate and w the crystal rotation rate, and is therefore the distance which the crystal grows during one rotation. The effective segregation coefficient, which determines the incorporation of impurities into the silicon lattice during the crystal growth process depends - as described above - on the actual growth rate v, and the thickness δ of the diffusion layer and the diffusion coefficient D of the impurities in the silicon melt. The crystal rotation in an asymmetric temperature crystal rotation field induces a periodic temperature variation, which causes a fluctuation of the thickness of the diffusion layer and thus a difference in the effective segregation coefficient of impurities into the crystal.

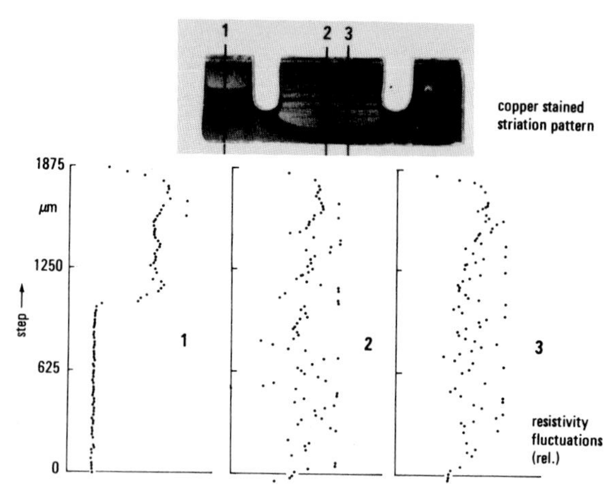

Fig. 13. Axial resistivity fluctuation (striations) of a lithium-drifted diode.

The variations in the concentration of striated impurities can gradually be reduced by choosing growth conditions which minimize fluctuations of v and δ.

Figure 14 shows the copper staining results of the (a) axial and (b) lateral cross-section of a 3" diameter diode which had been lithium drifted for 100 days (110°C, 500 V). It is well known that copper preferentially decorates n-type regions. Again, from top to bottom the n^+-layer (light), the lithium-drifted layer (i-layer) and the non-compensated p-layer are clearly revealed. Superimposed on the i-layer and the p-layer appears the typical resistivity pattern. The i-region is about 5 mm thick at the periphery of the crystal and only 1.5 mm in the center. The lateral cross-section shows that the impurity concentration also fluctuates and, in addition, the radial impurity distribution is not symmetrical. Similar distributions have alrady been found by Mühlbauer on silicon slices using the spreading resistance method. In this case, the lines were lines of constant resistance. The main cause for the fluctuations of the resistivity near the center was the <111> single crystal orientation. At the center of the growing interface a facet forms independent of the growth speed. Changes in the growth mechanism lead to an increased incorporation of dopant. This can result in a severe drop in the resistance at the center. Here, however, the striations are p^+-type, as can be seen in the modulation of the lithium-mobility in the lateral cross section.

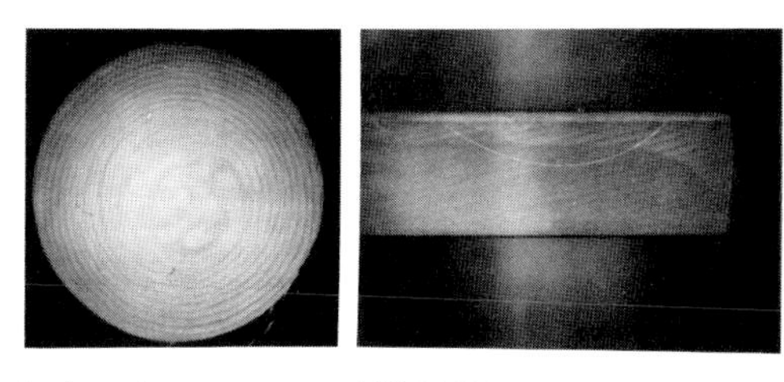

Fig. 14. Striation pattern of a 3" diam. 5 mm thick silicon diode after 100 days lithium drift at 110 °C and 500 V, revealed by copper staining.

The overall drift profile however can not be explained entirely by the influence of the striations; the cause may be due in part to a large radial variation in point defects, which have a lesser concentration at the perifery (due to out diffusion).

Swirl defects

In dislocation-free FZ silicon crystals microdefects arranged in a spiral or striated distribution in planes perpendicular or parallel to the growth direction, respectively, can often be revealed by preferential etching (Figure 15). At least two types of these microdefects - usually called swirl defects

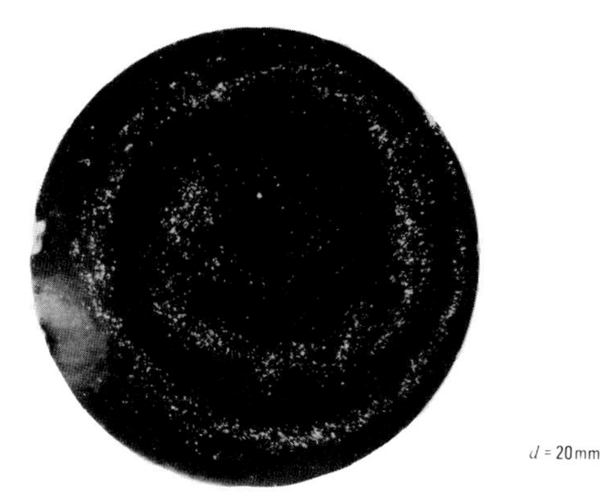

Fig. 15. Swirl pattern on the perpendicular cross section of a FZ crystal made visible by preferential etching.

due to the characteristic pattern - have been identified [20, 36]. Figure 16 shows A-swirl and B-swirl defects, revealed in lithium drift quality material. A- and B-swirl defects differ in size, concentration and spatial distribution. Apart from preferential etching swirl defects can be revealed by x-ray topography techniques in particular in combination with decoration methods [20, 36].

The microscopic structure of swirl defects has been unravelled by transmission electron microscopy (TEM), in particular by high-voltage TEM (HVEM). It

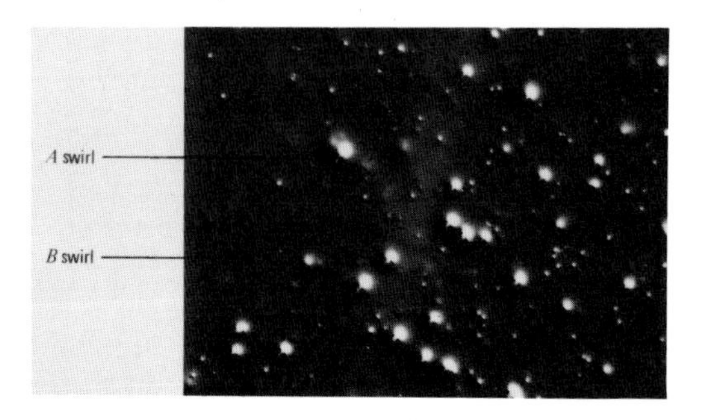

Fig. 16. Magnified part of swirl etch pattern displaying A swirl defects as larger black/white features (hillocks) and B swirl defects as smaller black/white dots (pits or hillcocks).

It was established that A-swirl defects consist of perfect dislocation loops of a typical size of about 1 μm [37]. They occur as single loops or complicated loop arrangements (Figure 17). Moreover, the analysis of the nature of these loops disclosed that they are of interstitial type [38, 39]. The nature of "clean" B-swirl defects could not be identified by TEM [39]. B-swirl defects decorated already in-situ during crystal growth by metal impurities are imaged as coherent or semi-coherent precipitates [38]. The results of TEM and etching indicate that the B-swirl defect themselves create a very small strain field and that their size is very likely in the order of tens of nanometers or less.

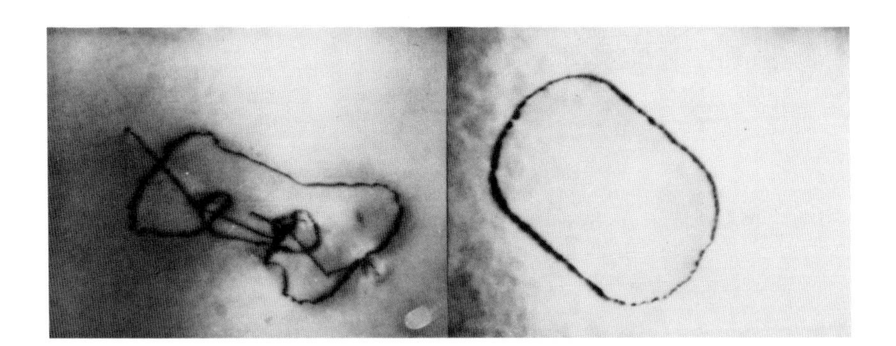

Fig. 17. HVEM (650 keV) micrographs of A swirl defects in FZ silicon

a) Single loop
b) Complex loop structure

The occurrence, concentration, size and distribution of swirl defects is strongly influenced by growth parameters and impurities. Fig. 18 illustrates the relationship between the concentration of A- and B-swirl defects and the growth rate for crystals of 25 mm diameter [20]. A- and B-swirl defects are not formed at $v \geq 5$ mm/min and $v \leq 0.2$ mm/min. With increasing crystal diameter the upper critical growth rate where swirl defects are eliminated decreases [40]. In 75 mm crystals swirl defects vanish already at $v \geq 3$ mm/min. The carbon concentration of the crystals strongly influences the concentration, size and occurrence of swirl defects, in particular of B-swirl defects [41]: With increasing carbon content the concentration of B-swirl defects and the critical growth rate for their elimination increases. The average concentration of A-swirl defects was found to be typically 10^6 to 10^7 cm^{-3}; the concentration of B-swirl defects ranges from 10^7 to 10^{11} cm^{-3}. A peripheral crystal region about 1 to 2 mm in width is depleted of swirl defects. Also around dislocations a zone of about 1 mm in width is free of swirl defects. By addition of hydrogen to the growth atmosphere the formation of swirl defects is prevented [20, 36]. Also doping of the silicon with phosphorus or oxygen exerts an inhibiting effect on swirl formation [40]. Cu decorates readily both A- and B-swirl defects, but Li decoration applies to A-swirl defects only.

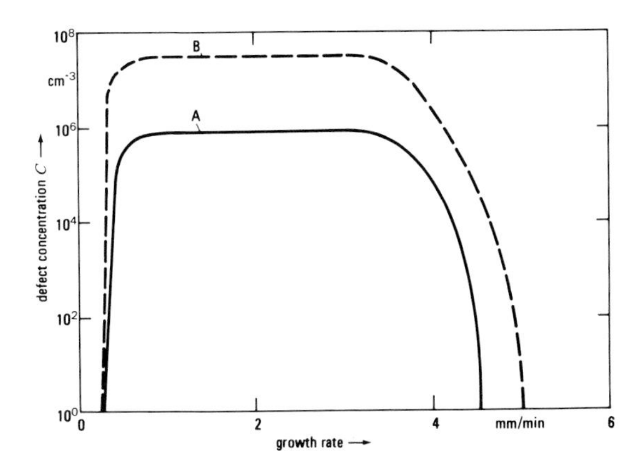

Fig. 18. Experimental concentration of A-type and B-type swirl defects as a function of growth rate (based on [54]).

In CZ silicon swirl defects of the A-, B- and C-type have very recently been discovered and studied [42, 43]. Several models for the formation of swirl defects in FZ silicon have been conceived [20, 36, 44, 45, 46] and reviewed recently [29, 47, 48, 49] . At present none of these models can explain convincingly all the experimental results. Common basis of all models is that a super-saturation of intrinsic thermal point defects (silicon self-interstitials, vacancies) is the driving force in the formation of swirl defects and that heterogeneous nucleation most likely involving carbon dominates. The models of Petroff and de Kock [44] and Föll et al [45] assume that swirl defects are formed by condensation of silicon interstitials and that the B-swirl defects are precursors of the A defects. B defects are postulated [45] as three-dimensional agglomerates of silicon interstitials and carbon atoms which collapse when exceeding a critical size into the interstitial-type dislocation loops (A-defects). Chikawa and Shirai [46, 48] proposed a model in which liquid drops are formed during the remelt cycle behind the solid liquid interface. During cooling of the crystal the drops convert to A and/or B defects. Van Vechten [47] conceived a model which tries to explain the formation of interstitial type dislocation loops by condensation of vacancies. According to a proposal of Hu [50] A defects are viewed as agglomerates of silicon interstitials and B defects as agglomerates of vacancies. Some recent results on diffusion experiments in silicon support the view that vacancies and silicon interstitials may coexist in silicon at elevated temperatures without significant mutual annihilation [51]. Very recently a model was developed by Voronkov, which is based on the coexistence of vacancies and self interstitials. In this model the recombination of vacancies and self interstitials is described as a function of crystal growth rate and temperature gradient. According this model some phenomena not understood up to now may be explained [57].

In FZ crystals with diameters above 45 mm a socalled etching depression in the center of crystals can occasionally be found after applying a usual polish-etch treatment [7, 52] . This defect seems to be related to B-swirl defects. It is also not detectable by TEM. The etching depression defect can readily be decorated by Cu and Au.

Swirl defects can exert a pronounced influence on the detector properties. Fong et al [54] very recently reported on low bias charge collection analysis of Si(Li) pin diodes prepared from large diameter (50 mm) FZ silicon crystals. They found poor charge collection in crystal regions revealing swirl defects after preferential etching. We have observed a correlation between poor performance of X-ray detectors in particular poor energy resolution and swirl defects. In Table IV the results obtained on 8 crystals supplied by two manufacterers are listed. Fig. 19 gives examples for the energy resolution in terms of FWHM of the ^{55}Fe line at 78 K. Detectors prepared from crystals containing swirl defects generally show poor energy resolution (> 170 eV), some even are characterized by reduced breakdown voltage. However, we obtained also poor detector performance with swirl-free material (72687) and, on the contrary also fairly

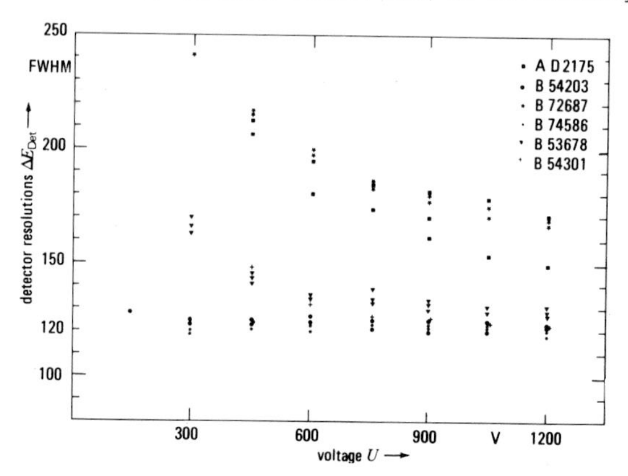

Fig. 19. Detector resolution ΔE_{Det} (FWHM) for ^{55}Fe-X-rays (5.9 keV) measured with 12 mm diam., 3 mm thick Si(Li) detectors, made from 6 different ingots.

good performance with crystals (1540 D, 54203, 54301) displaying swirl defects after preferential etching. From Table IV it is also evident, that all crystals of manufacturer A contained rather heavy swirl defect bands whereas those of manufacturer B apart from one exception (53678) display no or just weak swirl defect bands. This indicates that different crystal growth conditions are used by manufacturer A and B. From the striation and swirl pattern we surmise that the solid-liquid interface shape was flatter in the case of manufacturer A than for B.

TABLE IV
Comparison of detector properties and swirl defects in different silicon materials [15]

ingot	crystal-diameter mm	detector energy resolution for X-rays (78 K)	swirl–A–defects	solid / liquid interface
A 1537 D	31	bad	pronounced bands high density	very bent
B 54203	36	very good	weak	flat
B 53678	21	medium	pronounced bands high density	very bent
A 1540 D	30	medium –bad	pronounced bands in the center also A	slightly bent
B 54301	27	good at high voltages	weak	flat
A 2175 D	31	bad	pronounced bands	very bent
B 72687	37	bad	no bands	
B 74586	34	very good	swirl free	

The influence of the crystal growth conditions on the detector performance has recently been demonstrated by Fong et al [54]. These authors used 50 mm FZ silicon crystals for detectors grown by two different processes: In process I the molten zone and the melt flow pattern was distinctly larger than in process II. Si(Li) diodes prepared on process I silicon slices exhibited a relatively high and homogeneous charge collection at low bias, whereas those of process II silicon showed low and inhomogeneous charge collection across the slice diameter. In the case of process II crystals grown at high speed an etching depression ring found in the center corresponded to diode regions of low charge collection. Fong et al explain their results by perturbation of the nucleation process at the solid-liquid interface in process II crystals resulting in the creation of defect complexes similar to the early stages of swirl defect generation. These defects act as trapping centers for charge carriers. However, one of the process I crystals which had also an etching depression in the center gave good and homogeneous charge collection. Guislain et al [55] found a ten times faster precipitation of Li in swirl-free crystals than in crystals with swirl defects or dislocations. In the latter the Li precipitated preferentially at those defects. These authors also reported that the Li drift mobility in crystals with swirl defects was higher (1.5×10^{-10} cm^2/Vs) than in swirl-free crystals (0.2×10^{-10} cm^2/Vs). Guislain and De Laet [56] observed also that in some of their crystals the Li mobility at the periphery of the slice was about five times higher than at the center. We also found this effect in some of our crystals investigated whereas Fong et al [54] did not observe this effect. They determined Li mobilities of about 2.3×10^{-10} cm^2/Vs for various crystals with and without swirl defects and etching depression.

Guislain et al [55] tried to explain the retarted Li drift by a vacancy supersaturation which should exist in dislocation-free and swirl-free crystals except at the periphery. In crystals with dislocations or swirl defects those act as sinks for excess vacancies. The same applies to the crystal surface, thus a region at the periphery of about 2 mm is depleted of vacancies. Apart from the fact that in the meantime it was verified that at least A-swirl defects form by condensation of excess silicon self-interstitials (see above) the interpretation of the retarded Li drift in terms of a vacancy or a self-interstitial supersaturation is not convincing. Every crystal is depleted of vacancies and self-interstitials at the rim but only some crystals show the effect of Fig. 14. For many crystals which do not reveal any defects by preferential etching no retarded Li drift was established, even though in such crystals a supersaturation of vacancies and self-interstitial should be present by theory.

Thus at present the picture is still rather confusing. Therefore additional experiments are necessary in order to arrive at a better understanding of these material phenomena itself and their influence on detector properties.

ACKNOWLEDGEMENTS

The authors wish to thank Mr. B. Reiss and Mr. P. Steckel who prepared the lithium drifted detectors and made the copper staining and Mrs. Mylonas for swirl etching. The authors wish also to thank Dr. W. Keller and Dr. K. Reuschel for reading the manuscript and for their valuable criticism.

REFERENCES

1. E. Spenke, W. Heywang, Siemens Review, 48/1 (1981) 4.

2. P. A. Glasow, IEEE Trans. on Nucl. Science NS 29-3 (1982) 1159.

3. R. N. Hall, T. I. Soltys, IEEE Trans. on Nucl. Science NS 18-1 (1971) 160.

4. W. L. Hansen, Nucl. Instr. Methods 94, (1971) 377.

5. H. F. Hadamowsky, "Halbleiterwerkstoffe", VEB Deutscher Verlag für Grundstoffindustrie, Leipzig 1972.

6. W. G. Pfann, "Zone Melting", John Wiley & Sons, Inc. New York · London · Sydney, 1966.

7. W. Keller, A. Mühlbauer, "Floating-Zone Silicon", Marcel Dekker, Inc., 1981 New York and Basel.

8. A. Yusa, Y. Yatsurugi, T. Takaishi, Journ. Electrochem. Soc. 122, (1975) 1700.

9. F. Shiraishi, M. Hosol, Y. Takami, Ohsawa, IEEE Trans. on Nucl. Science, 29-1 (1982) 775.

10. W. Keller, H. Stutt, Z. Feinwerktechnik 75 (1971) 207.

11. W. C. Dash, J. Appl. Phys. 30 (1959) 459
W. C. Dash, J. Appl. Phys. 31 (1960) 736.

12. G. Ziegler, Internal Siemensreport 1-108, Munich, 1960
G. Ziegler, Z. Naturforsch., (1961) 219.

13. W. Keller, DBP 2.358.300, filed Nov. 22, 1973, patented July 20, 1978;

USP 3,923,468, filed Nov. 20, 1974, patented Dec. 2, 1975;
USP 3,961,906, filed May 27, 1975, patented June 8, 1976.

14. H. Herzer, and H. Zauhar, USP 4,060,392, filed May 5, 1976, patented Nov. 29, 1977.

15. B. Kolbesen, in "Large Scale Integrated Circuit Technology State of the Art and Prospects" (1982) 33, Editors: Leo Esaki and Giovanni Soncini Martinus Nijhoff Publisher Ten Hague, Netherland.

16. K. Lark-Horowitz, in Proceedings of the Conference at the University Reading, Butterworth, London, 1951.

17. M. S. Schnöller, W. E. Haas, J. Electronic Mater. 5 (1976) 57.

18. W. Maenhaut and J. P. op de Beck, J. Radioanal. Chem. 5, (1970) 115.

19. C. Kim, H. Kim, A. Yusa, S. Miki, K. Husimi, S. Ohkawa, Y. Fuchi, IEEE Trans. on Nucl. Science, NS 26-1, (1979) 292.

20. A. J. R. de Kock, in Handbook on Semiconductors, ed. T. S. Moss, 3, ed. S. P. Keller, North Holland, Amsterdam (1980) 247.

21. B. O. Kolbesen, I. Simposio Brasileiro de Microelectronica, 9 A 11 de Septembro de 1981, Sao Paulo - Brasil.

22. B. O. Kolbesen, A. Mühlbauer, Solid State Electronics, 25-8 (1982) 759.

23. K. E. Benson, W. Lin, E. P. Martin, in Semiconductor Silicon, eds. H. R. Huff, R. I. Kriegler and Y. Takeishi, The Electrochemical Soc., Pennington, N.Y. (1981) 33.

24. I. A. Baker, ref. [18] p. 566.

25. B. O. Kolbesen, Applied Physics Letters, Vol 27-6, (1975) 353.

26. J. Kestler, B. Reiß, W. Czulius, P. Glasow, to be published.

27. H. J. Guislain, W. K. Schoenmakers, L. H. Delaet, Nucl. Instr. and Methods 101 (1972) 1.

28. W. Frank, Festkörperprobleme XXI (Advances in Solid State Physics), J. Trensch (ed), Vieweg, Braunschweig (1981) 221.

29. A. J. R. de Kock, Defects in Semiconductors, ed. J. Narayem and T. Y. Tan, North Holland, Amsterdam (1981) 309.

30. A. Seeger, H. Föll and W. Frank, Radiation Effects in Semiconductors, Inst. Phys. Conf. Ser. 31, (1977) 12.

31. H. Chisaka, M. Masuda, Nucl. Instr. and Methods 98 (1972) 255.

32. K. R. Mayer, J. Electrochemical Soc., 120 (1973) 1780.

33. A. J. R. de Kock, S. D. Ferris, L. C. Kimerling, H. J. Leamy, Appl. Phys. Letters 27 (1975) 313.

34. J. Burtscher, Scientific Principles of Semicond. Techn., Proc. European Summer School, ed. H. Weiß, (1974) 63.

35. H. Herrmann, H. Herzers and E. Sirtl, Festkörperprob. XIV (Adv. in Solid State Phys.) H. J. Queisser ed, Perg./Vieweg, Braunschweig (1975) 279.

36. A. J. R. de Kock, Philips Res. Rept. Suppl. No. 1 (1973).

37. L. I. Bernewitz, B. O. Kolbesen, K. R. Mayer and G. E. Schuh, Appl. Phys. Lett. 25, 277 (1974).

38. P. M. Petroff and A. J. R. de Kock, J. Cryst. Growth 30, 117 (1975).

39. H. Föll and B. O. Kolbesen, Appl. Phys. 8, 319 (1975).

40. T. Abe, H. Harada and J. Chikawa, Physica B, (Proc. 12th Int. Conf. Defects in Semiconductors, Amsterdam, 1982) in print.

41. H. Föll, U. Gösele and B. O. Kolbesen, in Semi. Silic. 1977, ed. by H. R. Huff and E. Sirtl, The Electrochem. Soc., Princeton, N.J. p. 565.

42. A. J. R. de Kock, W. T. Stacy and W. M. van de Wijgert, Appl. Phys. Lett. 34, 611 (1979).

43. A. J. R. de Kock and W. M. van de Wijgert, J. Cryst. Growth 49, 718 (1980).

44. P. M. Petroff and A. J. R. de Kock, J. Cryst. Growth 35, 4 (1976).

45. H. Föll, U. Gösele and B. O. Kolbesen, J. Crystal Growth 40, 90 (1977).

46. J. Chikawa and S. Shirai, J. Crystal Growth 39, 328 (1977).

47. J. A. van Vechten, Phys. Rev. B-17, 3197 (1978).

48. J. Chikawa and S. Shirai, Jap. J. Appl. Phys. 5. 153 (1979).

49. H. Föll, U. Gösele and B. O. Kolbesen, J. Cryst. Growth 52, 907 (1981).

50. S. M. Hu, J. Vacuum, Sci, Techn. 14, 17 (1977).

51. D. A. Antoniadis, J. Electrochem. Soc. 129, 1093 (1982).

52. W. Keller and A. Mühlbauer, Inst. Phys. Conf. Ser. 23, 538 (1975).

53. P. J. Roksnoer, W. J. Bartels and C. W. Bulle, J. Cryst. Growth 35, (1975) 245.

54. A. Fong, J. T. Walton, E. E. Haller, H. A. Sommer, J. Guldberg, Nucl. Instr. and Methods 199, (1982) 623.

55. H. J. Guislain, W. K. Schoenmakers, L. H. Delaet, Nucl. Instr. and Methods 101 (1972) 1.

56. H. J. Guislain and L. H. Delaet, IEEE Trans. on Nucl. Science, NS 19/1 (1972) 323.

57. V. V. Voronkov, J. of Cryst. Growth, Bd 59 (1982) 625 - 643.

FABRICATION OF ULTRA HIGH PURITY SILICON SINGLE CRYSTALS

D. ITOH, *S. KAWAMOTO, S.MIKI, I. NAMBA AND Y. YATSURUGI
Komatsu Electronic Metals Co., Ltd., 2612 Shinomiya, Hiratsuka, Japan 254
*75-52 Kessel St., Forest Hills, N.Y. 11375, U.S.A.

ABSTRACT

High resistivity silicon single crystals for radiation detectors were grown by the floating-zone (FZ) method with and without one-pass zone refining of ultra pure starting material obtained by the decomposition of monosilane gas. In this method, boron is removed from monosilane during gas generation. The main residual impurity is phosphorus from phosphine which is removed with zeolite A after distillation of monosilane. Photoluminescence analysis and an improved 4-point probe resistivity measuring method were used to evaluate the material.

INTRODUCTION

Many efforts have been made to obtain high resistivity silicon single crystals for radiation detectors, using polysilicon rods produced from trichlorosilane [1], [2], [3]. Though impurities other than boron can be removed by the vacuum zone-refining method, it is almost impossible to remove boron by this method. Therefore, it appeared difficult to obtain high resistivity silicon reliably on a mass production scale by the trichlorosilane method. Polysilicon rods were produced by purifying and decomposing monosilane gas, which has the characteristics listed below. The polysilicon rods are then converted into single crystals. The advantages of obtaining silicon from monosilane decomposition are:

(a) During monosilane decomposition to silicon and hydrogen no corrosive gas such as chlorine is generated as is the case when trichlorosilane is used.
(b) The decomposition temperature of monosilane is about 800°C, several hundred degrees centigrade lower than the reducing temperature in the trichlorosilane method.
(c) In the generation process of monosilane, boron, which is almost impossible to remove by zone-refining in vacuum, can be removed from the monosilane by the following chemical reaction:

$$\equiv B + NH_3 \dashrightarrow \equiv B:NH_3 \qquad (1)$$

Further, the concentration of boron in the monosilane gas estimated from the silicon product is below 0.005 ppba, which is considerably lower than that of trichlorosilane.

PREPARATION OF MATERIALS

Monosilane gas

In the production process, monosilane is generated by the following chemical reaction between magnesium silicide and ammonium chloride in liquid ammonia:

$$Mg_2Si + 4NH_4Cl + 8NH_3 \dashrightarrow SiH_4 + 2(MgCl_2 \cdot 6NH_3) \qquad (2)$$

As mentioned previously, boron hydride is removed in the following chemical reaction:

$$B_2H_6 + 2NH_3 \dashrightarrow B_2H_6 \cdot 2NH_3\downarrow \quad (3)$$

The phosphine concentration is reduced to about 0.05 ppba after distillation. Further, to remove the phosphine, adsorptive purification is used with ion-exchanged zeolite [4]. The zeolite is pretreated by ion-exchanging alkaline ions in molecular sieve 3A with zinc ions. The concentration of phosphine after the adsorptive purification is below 0.01 ppba.

Polysilicon

Polysilicon rods were produced by decomposing the purified monosilane gas on to silicon wires. The decomposition temperature was between 810°C and 860°C, and rods with diameters from 70 to 80 mm were produced. The n-type silicon wires themselves were produced by the pulling method, yielding resistivities of about 100 ohmcm.

Processing

The core wire of the polysilicon rod was cut out and disposed. Only the grown layers of polysilicon were furnished to single crystal manufacturing after being ground to cylindrically shaped rods, 30-35 mm in diameter and 500-600 mm in length. Two such rods can be extracted from a single 70-80 mm diameter polysilicon rod. To remove contamination, the surfaces of the rods were etched with electronic grade HF-HNO_3 solution and rinsed with deionized water having a resistivity of 15-18 MΩcm. Special attention was paid to prevent external contamination.

Zone-Refining and Crystallization

A special floating-zone (FZ) refining machine has been developed for the fabrication of high resistivity silicon. A vacuum system composed of a rotary roughing pump trapped by a sorption pump and followed by a cryo pump produces a clean vacuum in the 2×10^{-5} torr range in the zone refiner. Argon gas of 99.9% purity was used as a backfill during crystal growth. A commercially available r.f. oscillator powered a one-turn r.f. coil which together with a thermal reflector produced the molten zone. Prior to use, the FZ machine was cleaned by several passes in vacuum of pure polysilicon rods. This conditioning is very important for outgassing and coating the inside wall of the machine and surfaces of the zone refiner, including the r.f. coil, with a thin film of clean silicon.

Single crystals with the resistivities between 40-60 KΩcm were obtained only by crystallization in argon gas atmosphere without zone-refining in vacuum. Silicon with higher resistivity was obtained first by zone-refining in vacuum (refining speed: about 5 mm/min, zone length: about 15 mm) to evaporate some small amount of phosphorus and then crystallizing in the argon gas atmosphere. P-type single crystals with resistivity above 30 KΩ-cm were obtained by crystallization after one pass zone-refining (refining speed: about 1 mm/min and zone length: about 30 mm).

Evaluation Method

Resistivity was measured by both 2-point probe and 4-point probe methods. The measurement of high resistivity silicon is very difficult due to the effect of contact resistance between the probe and specimen which can be minimized by using a high impedance operational amplifier in the voltage measuring circuit. Concentrations of shallow level impurities, such as phosphorus, boron, arsenic, etc. were measured by the photoluminescence (PL) analysis method as improved by Tajima [5]. The data derived by this method are more accurate and easily obtained than that from radioactivation analysis or Hall coefficient measurement. An etched surface of the specimen was measured by the PL method at liquid helium temperature. Luminescence was excited by 514.5 nm light from an Ar-ion laser which was measured by a SPEX 1701 grating monochromator and HTV 7102 photomultiplier cooled by dry ice. The resultant PL signals were transmitted to a computer for data processing. Figure 1 shows the block diagram of the apparatus and Figure 2 shows the calibration curves for phosphorus and boron. The detection limit of this method for phosphorus is 0.001 ppba, for boron, 0.002 ppba, and for other shallow level impurities 0.002 ppba. The lifetime of the specimen was measured by the photoconductive decay method and the conductive type was determined by the thermal electromotive method. An attempt was made to measure the concentrations of carbon and oxygen by infrared spectroscopic analysis, however the concentrations were below detectable limits at room temperature.

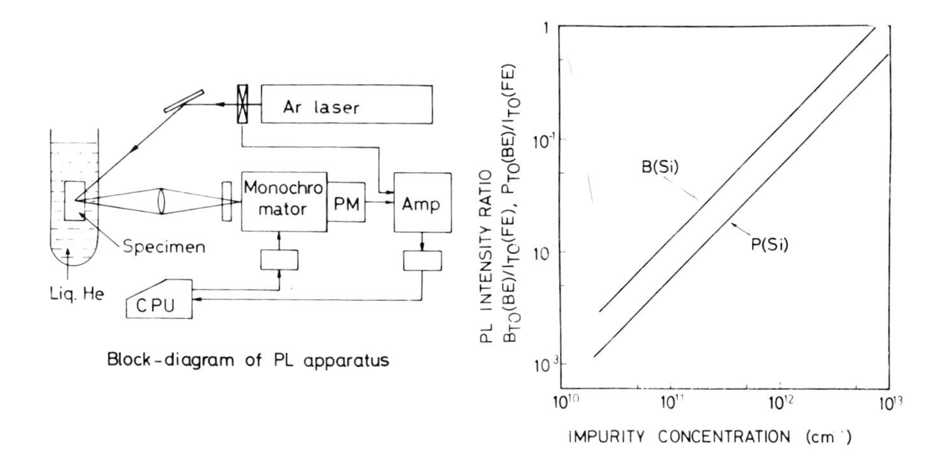

Fig. 1. Block diagram of photoluminescense evaluation apparatus (PL).

Fig. 2. Calibration curves for the photoluminescense method for B and P in Si. The PL intensity ratio between intrinsic and extrinsic components is shown for the respective impurities.

RESULTS AND DISCUSSION

As the resistivity of high purity silicon is generally measured by widely-spaced 2-point probes, only the average resistivity of a given region (in the cross-sectional and longitudinal direction) can be measured. Narrow spaced 4-point and 2-point probe methods have been used to obtain precise local resistivity fluctuation of a single crystal. However, the measuring procedure of the 4-point probe method is very difficult resulting often in measurement errors.

In order to minimize these errors, we have studied the measuring technique. The results in Figure 3 show that for higher resistivities, the ohmic regime extends to higher electric fields in the higher resistivity specimens. The drop in resistivity at high fields is probably caused by minority carrier injection. From these results, the proper electric field for high resistivity

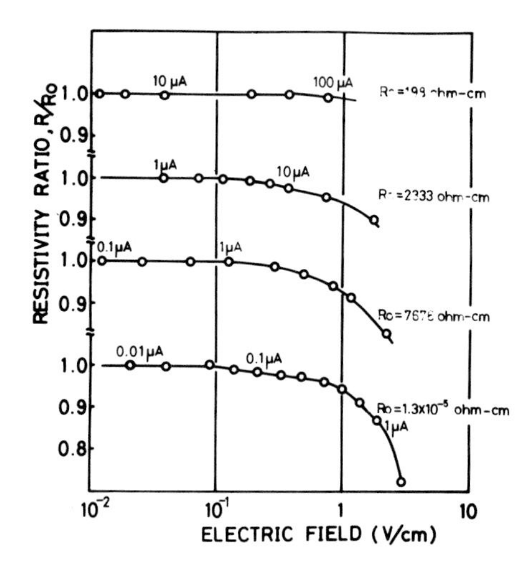

Fig. 3. Relation between resistivity and electric field in 4-point probe measurement.

measurement for the 4-point probe method has beenobtained. That is, for the resistivity measurement to be truly ohmic, the electric field should be lower than 0.1 V/cm. Thus, the current for resistivity measurement by the 4-point probe method (1 mm probe interval and voltge lower than 0.01 V) is obtained from the following equation:

$$R = 2\pi s\ V/I$$

where s is the probe interval. The current is calculated by:

$$I = 1/(160 \times R).$$

In our measurement of p-type high resistivity silicon, specimens were prepared without conventional lapping, hydro-sandblasting and etching. Inversion layers (n-type layer) are formed on the surface of the specimen by such preparations. We prepared the specimen by lapping with oil, and then washing with trichloroethylene and acetone. There was not much difference between data obtained with the 4-point probe and 2-point probes.

Figures 4 and 5 show the PL spectra of n-type and p-type high resistivity silicon, respectively. The PL spectra of the n-type specimens (Figure 4a) prepared by crystallization in an argon gas atmosphere, having the polysilicon produced from monosilane gas, does not contain impurities other than phosphorus and boron, such as arsenic, gallium, etc. Further, it is noted that the resistivity of the polysilicon is about 50 KΩcm (n-type), and about 0.004

ppba boron and 0.006 ppba phosphorus are contained in it. Figure 4(b) shows the PL spectra of the n-type silicon produced by vacuum zone-refining and crystallization. High resistivity of n-type silicon was achieved by the slight difference between donors and acceptors, and $N_D/N_A > 0.3$. The cross-sectional resistivity deviation is between 30% and 50%.

Figure 5 shows the PL spectra of the p-type high resistivity silicon produced by zone-refining and crystallization. By increasing the phosphorus evaporation rate, its concentration in the silicon can be reduced to 1/4 to 1/6 of that in the starting polysilicon material; the ratio N_D/N_A can also be reduced to below 0.2. The cross-sectional resistivity deviation can be reduced to 20%. Boron concentration ranged from 0.002 ppba to 0.005 ppba and the maximum resistivity achieved was 150 KΩcm. Various characteristics of the high purity silicon are summarized in Table I.

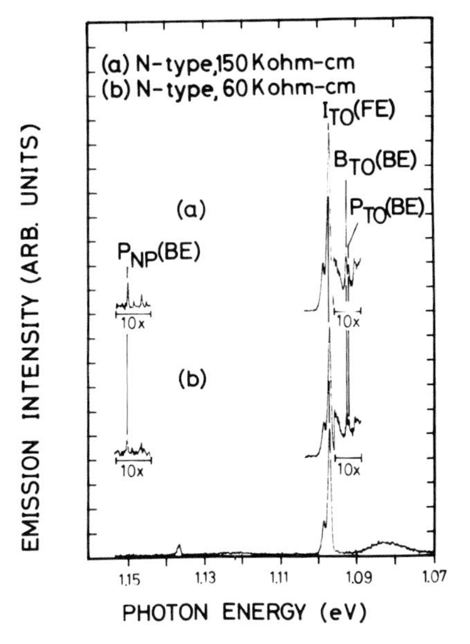

Fig. 4. PL spectra of N-type high resistivity silicon.

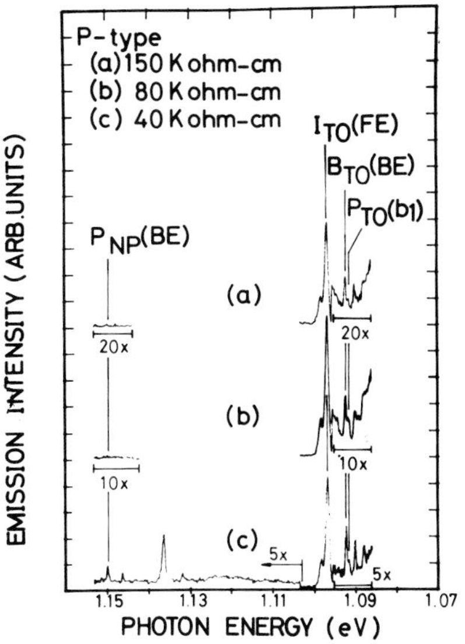

Fig. 5. PL spectra of P-type high resistivity silicon.

TABLE I
Characteristics of high purity silicon

Item	Type	Resistivity (K ohm-cm)	N_D(P) (x 10^{11})	N_A(B) (x 10^{11})	Lifetime (m sec)	Fabrication Method
Fig. 4(a)	N	150	1.5	1.2	3.8	FZ in Ar after in vac.
Fig. 4(b)	N	60	2.9	2.2	4.2	FZ in Ar
Fig. 5(a)	P	150	0.2	0.2	3.4	FZ in AR after in vac.
Fig. 5(b)	P	80	0.3	2.3	3.8	FZ in AR after in vac.
Fig. 5(c)	P	40	2.3	2.0	4.0	FZ in AR after in vac.

A surface barrier detector was fabricated from high purity silicon, 50 KΩcm N-type. Figure 6 shows a representative spectrum measured at 77K for X rays from ^{241}Am. The total resolution (FWHM) was 703 eV at 17.74 κeV with the 2200 V bias on this detector.

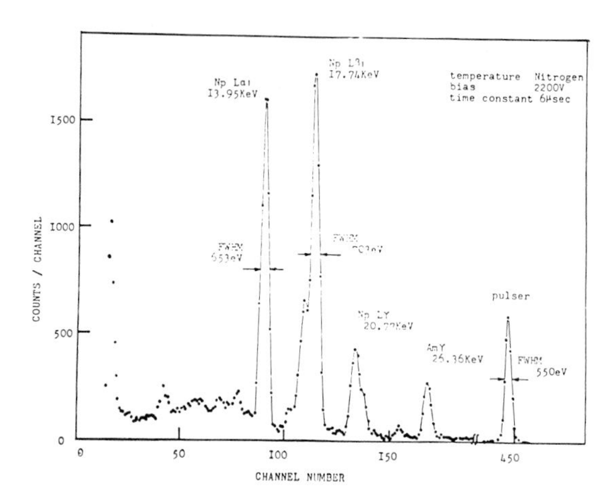

Fig. 6. Spectrum of X rays from ^{241}Am obtained by a high resistivity silicon detector with the side groove.

CONCLUSION

Using polysilicon materials produced from highly refined monosilane gas, we have succeeded to simply and stably produce single crystal silicon of detector grade by controlling processing. Careful evaluation using the PL method and improved resistivity measurements are necessary to detect and minimize contamination.

ACKNOWLEDGMENTS

The authors would like to express sincere appreciation to K. Fushimi of the University of Tokyo and C. Kim of Korea University who fabricated and evaluated the solid state detectors. Further, the authors express sincere appreciation to M. Tajima of Electrotechnical Laboratory for his guidance and advice on the PL evaluation.

REFERENCES

1. A. Hoffmann, K.Reuschel and H. Rupprecht, J. Phys. Chem. Solids, 11, 284 (1954).

2. T. G. Digges, Jr. and C. L. Yaws, J. Electrochem.Soc., 121, 1221 (1977).

3. E. L. Kern, L. S. Yaggy and J. A. Baker, Semiconductor Silicon 1977 (Proceedings of the Third International Symposium on Silicon Materials Science and Technology), 52, (1977).

4. A. Yusa, Y. Yatsurugi and T. Takaishi, J. Electrochem. Soc., 122, 1700 (1975).

5. M.Tajima, A. Yusa and T. Abe, Jpn. J. Appl. Phys., 19, 631 (1980).

THE CHARACTERIZATION OF GERMANIUM AND SILICON FOR NUCLEAR RADIATION DETECTORS.*

L. S. DARKEN, Solid State Division, Oak Ridge National Laboratory, P.O. Box X, Oak Ridge, Tennessee 37830.

ABSTRACT

Semiconductor nuclear radiation detectors require deep depletion depths (0.03-3.0 cm) and effective charge collection distances which are several times longer than these depletion depths. These requirements place stringent limitations on the net electrically active impurity concentration, and on the concentration of deep centers which can trap carriers generated by the incident nuclear radiation. This need for extremely pure material distinguishes the interests and efforts of the semiconductor detector community from the rest of the semiconductor community. This paper reviews the characterization of shallow-level, deep-level, neutral, and extended defects in germanium and silicon for nuclear radiation detectors. Photothermal ionization spectroscopy has been used extensively to identify the residual hydrogenic impurities in high-purity ($|N_A-N_D| \approx 10^{10}-10^{11} cm^{-3}$) germanium and silicon. Deep level transient spectroscopy has been effectively used to detect and identify deeper levels in high-purity germanium. Residual neutral defects are not necessarily passive: they may complex to form deep or shallow levels, they may precipitate, or they may act as nucleation sites for precipitation. The properties of extended defects (dislocations, lineage, inclusions, precipitates) and their effects on device performance are fundamentally less well understood, as the origin of the electrical activity of these defects is uncertain. It has been found in numerous instances that chemical interactions among defects are important even in these high-purity semiconductors.

INTRODUCTION

The purpose of this paper is to review the current status of the characterization of the bulk properties of germanium and silicon for nuclear radiation detectors. Consideration will be restricted to material for high-purity rather than lithium-drifted detectors. Since the purity required ($|N_A-N_D| \approx 10^{10}-10^{11}$ cm^{-3}) is difficult to maintain consistently in a crystal growth process, routine measurement of the net electrically active impurity concentration is essential. In addition to predicting the immediate suitability of the semiconductor material for detector fabrication, characterization efforts are also directed toward a basic understanding of material properties and how they relate to crystal growth and device performance. The techniques used and results obtained in pursuit of both these objectives will be reviewed.

*Research sponsored by the Division of Materials Sciences, U.S. Department of Energy under contract W-7405-eng-26 with Union Carbide Corporation.

For the identification of residual shallow levels, photothermal ionization spectroscopy (PTIS) [1] has provided unequaled resolution and sensitivity. The most common shallow-level impurities found by PTIS are aluminum, boron, and phosphorus in germanium, and boron and phosphorus in silicon. Deep level transient spectroscopy (DLTS) [2] has likewise advanced the identification of residual deep levels in high-purity semiconductors. The poor resolution of some coaxial germanium detectors has been correlated with the concentration of copper-related deep levels as determined by DLTS [3]. Efforts to identify and quantify the residual neutral defects have been motivated by the reactivity of these elements. The concentrations in germanium of the normally neutral impurities carbon [4] and hydrogen [5] have been determined by incorporating radioactive isotopes of these elements during crystal growth, fabricating detectors for this material, and measuring the count rate of the beta particles generated during the nuclear decay. The lithium precipitation technique for measuring oxygen content in germanium is presented and analyzed with respect to relevant mass action equations, and the existence of an upper limit to its range of applicability is demonstrated. Extended defects, such as dislocations and particulates, can also either directly or indirectly affect electrical properties. There is neither an accepted microscopic theory for the electronic properties of these defects, nor much experimental data directly relevant to high-purity semiconductors for nuclear detectors. However, some interesting effects have been observed and some tolerance limits for acceptable detector resolution and leakage current have been proposed.

The paper is organized to cover the topics of shallow levels, deep levels, neutral defects, and extended defects in separate sections. This is also roughly in descending order of our basic understanding of these defects and how they influence detector performance. It is the author's intent that this paper serve as both a review and guide to the literature, as the breadth of scope necessarily precludes a thorough treatment.

SHALLOW LEVELS

The Group III acceptors and the Group V donors, as well as lithium, contribute shallow levels in germanium and silicon. In addition, particularly in germanium, numerous shallow-level complexes have been observed [6-9] and some identified. Shallow levels are sometimes called hydrogenic because their ionization energy is roughly approximated by the ionization energy of the hydrogen atom modified by the bulk dielectric constant of the semiconductor and the effective mass of the bound carrier. Usually in high-purity semiconductors for nuclear radiation detectors there are more of these shallow-level centers than deep-level centers. Therefore, charge transport in thermal equilibrium and the space charge density in the depletion region are dominated by the shallow levels. Transport techniques are routinely used to determine the net electrically active impurity concentration. In production, this is usually the only electrical characterization procedure between crystal growth and detector fabrication. Optical techniques, however, are used to determine the exact energy of the shallow levels and to thus identify the defects responsible.

Transport Techniques

A four-lead conductivity measurement can be performed on the as-grown or cut ingot with minimal sample preparation. Conductivity is measured using either a four-point probe or by passing current axially through the crystal and measuring the voltage along the side. The conductivity, σ, is proportional to the carrier concentration (n for electron concentration, p for hole concentration):

$$\sigma_n = ne\mu_n \qquad \sigma_p = pe\mu_p \quad , \tag{1}$$

where μ_n and μ_p are the carrier mobilities. Germanium is usually measured at 77 K ($\mu_n = \mu_p = 45,000$ V-sec/cm^2) and silicon, at room temperature ($\mu_n = 1500$, $\mu_p = 600$ V-sec/cm^2). Obtaining low impedance contacts, which are necessary for accurate transport measurements, is more of a problem on high-purity silicon at 300 K than on germanium at 77 K. Therefore, the net impurity concentration of a high-purity silicon ingot is frequently not known with certainty until a detector is fabricated and a capacitance-voltage curve obtained.

The carrier concentration can also be determined by the Hall effect coefficient R_H, which is usually measured in the van der Pauw configuration [10] on slices. The concentration as inferred from the Hall coefficient is not dependent on an assumed mobility, and the technique is fairly insensitive to radial concentration gradients. Carrier concentration p (n if R_H is negative) is related to R_H by the expression

$$p = \frac{\theta}{eR_H} \quad , \tag{2}$$

where θ is a factor near unity, which depends on the band structure of the carrier, orientation of the sample, temperature, and magnetic field strength [11]. The concentration of shallow-level centers can be distinguished from that of deep-level centers by measuring R_H as a function of temperature. The logarithm of the hole concentration p is plotted against the reciprocal temperature in Fig. 1 for two high-purity germanium samples: one dislocated, the other dislocation free. The dislocated sample shows classical one level behavior: an intrinsic region, an exhaustion region where the hole concentration is determined by the net concentration of shallow levels, and a freeze-out region where the holes are becoming bound to the shallow-level acceptors. The

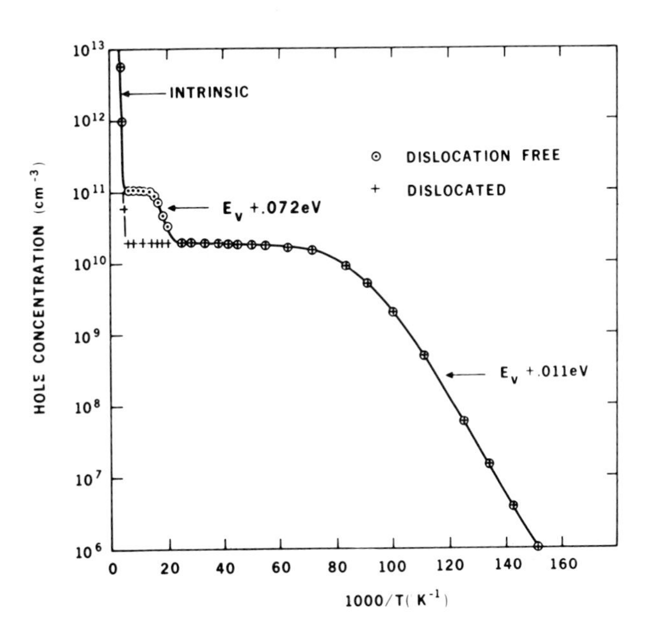

Fig. 1. Arrhenius plot of the free hole concentration p (log p versus 1000/T). The dislocation-free sample contains an acceptor level at E_V + 72 meV (Ref. 7).

dislocation-free sample has deeper levels which freeze out at a higher temperature than the shallow levels. These deep levels will be discussed further in the following section. The degree of compensation can also be inferred from analysis of R_H vs 1/T data.

At the normal operating temperature (~300 K for silicon, ~90 K for germanium) these shallow levels do not trap the carriers generated by ionizing radiation long enough to degrade the energy resolution of the detector; only their net concentration affects device performance through the required depletion voltage and the spatial dependence of the electric field in the detector. However, the development and control of high-purity crystal growth is guided by the identification of the residual shallow-level impurities. Ionization energies of these residual impurities can be determined in principle from R_H vs 1/T data, but in some cases this technique lacks the energy resolution to distinguish between, or determine the relative abundance of, energetically close levels. While Group III acceptors in silicon may be identified by measurement of R_H vs 1/T, Group III acceptors in germanium require a technique with better energy resolution.

Optical techniques

Optical techniques have been the most successful in identifying shallow-level species. Photoluminescence [12] has recently been developed to the sensitivity where it can be used on high-purity silicon. Radiative recombination of excitons bound to neutral donors and acceptors produces near-gap luminescence spectra characteristic of the center. The intensity of these features can be compared to the intensity of phonon-assisted exciton recombination lines. Photoluminescence spectroscopy has the advantage of simultaneously identifying and measuring the concentration of both donors and acceptors on small samples without requiring contacts. The technique has been calibrated for silicon and has a sensitivity near the range for detector-grade silicon (0.5×10^{11} cm^{-3} and 1.0×10^{11} cm^{-3} for phosphorus and boron, respectively) [13], and has proved useful in high-purity silicon growth [14].

Photothermal ionization spectroscopy (PTIS) is uniquely suited by its excellent energy resolution and high signal/noise ratio for the identification of residual shallow-level impurities in high-purity semiconductors. PTIS has been extensively used in the development of high-purity germanium [7-9,15,16] and was instrumental in focusing attention on residual chemical impurities [15,16]. In PTIS an electrically active center in the neutral state is photoconductively detected through a two-step ionization process: The electron (hole) in its ground state is raised to a bound excited state by infrared photon absorption and is photothermally ionized, if it escapes to the conduction (valence) band by phonon absorption before returning to the ground state. The process is shown schematically in Fig. 2 for holes in germanium. The sample is held at a temperature low enough to ensure population of the ground state, but high enough that holes (electrons) optically excited to higher bound states will be promoted to the valence (conduction) band. These photothermally generated holes (electrons) are then detected photoconductively.

The spectrum of a p-type high-purity germanium sample is shown in Fig. 3. Each acceptor contributes a series of sharp photoconductive responses corresponding to allowed ground state to excited state optical transitions. For concentrations of the compensating centers less than 10^{13} cm^{-3}, Stark effect broadening of linewidth due to inhomogeneities in the electric field produced by ionized centers is negligible, and linewidth broadening due to inhomogeneous strain is usually predominant. The broader direct ionization continuum is at a higher energy. Since the excited states observed have a vanishing wave function at the core (by the hydrogen atom approximation), the location of the excited states with respect to the valence band does not depend on the specific center involved. Thus the internal spacing of the transition series is the

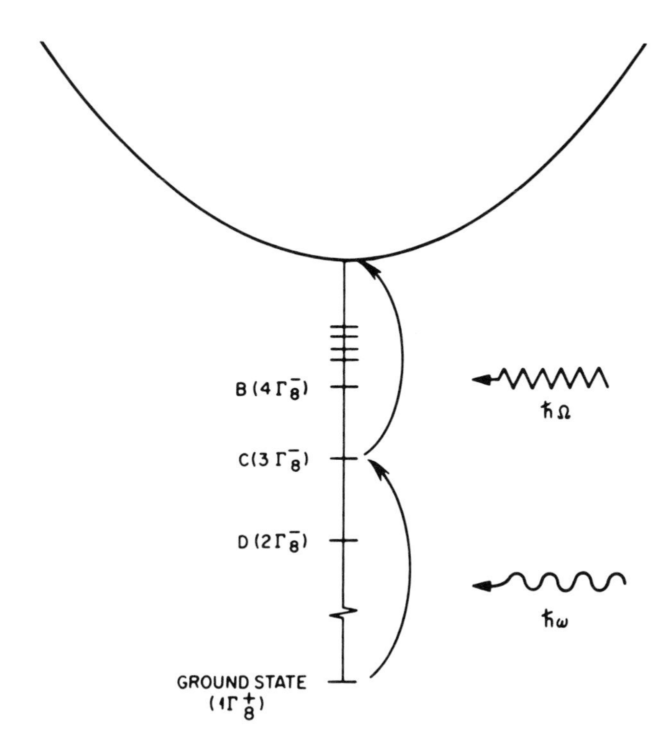

Fig. 2. Schematic representation of the photothermal ionization process in p-type germanium. The figure is drawn so that positive hole energy is upward. The strongest transitions are labeled and the identification of the states is given in parentheses (Ref.9).

same for all the acceptors shown. The residual acceptors in Fig. 3 are identified as boron, aluminum, and gallium. The gallium was intentionally added as a reference concentration in this particular crystal. This sample was simultaneously illuminated with band-edge light, so that the donors were also neutralized by the minority-carrier electrons generated. The principle residual donor was phosphorus. The photothermal ionization of the compensating donors contributed negatively to the conductivity. The polarity of the conductivity response from compensating centers will be discussed in more detail shortly.

In addition to the excellent energy resolution obtainable, PTIS retains a high signal-to-noise ratio at dilute concentrations of the shallow impurities. The photoconductive change in conductivity $\Delta\sigma$ is given by

$$\Delta\sigma = \mu e G \tau_{relax} \quad (3)$$

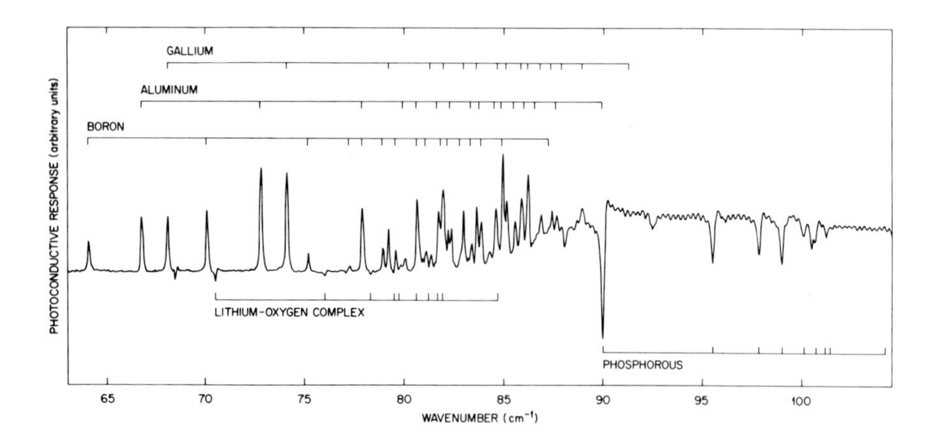

Fig. 3. Photothermal ionization spectrum of a p-type germanium sample illuminated with band-edge light so that some donors are also neutral. Donor peaks are negative. $|N_A - N_D| = 5\times10^{10}\ cm^{-3}$ (Ref. 9)

where G is the rate at which neutral centers are photothermally ionized and τ_{relax} is the relaxation time for the excess photothermally generated carriers to be recaptured by shallow levels. Since G is directionally proportional to the density of neutral centers and τ_{relax} is inversely proportional to the density of ionized centers available for recapturing the photothermally generated carriers, $\Delta\sigma$ (at constant T) is roughly independent of the absolute concentration of the residual shallow impurities [17].

When more than one shallow level impurity is present, the intensity ratio of corresponding transitions depends on the concentration ratio of the defects in the neutral charge state. In germanium, where the binding energies are fairly close, temperature-dependent statistical effects in the occupation of different shallow-level acceptors have been observed and used to obtain accurate concentration ratios [9]. The distribution of acceptors in a high-purity germanium crystal as determined by PTIS, Maxwell-Boltzmann statistics, and doping with a known quantity of gallium is shown in Fig. 4. This crystal was grown from an SiO_2 crucible under an atmosphere of hydrogen. The abscissa in Fig. 4 was chosen so that impurities which segregated according to the normal distribution law during solidification will be plotted linearly. The position (normalized by mass) of the sample in the crystal is given directly at the top of the figure. Gallium segregates as expected: $k_{eq}=0.10$, $k_{eff}=0.12$. However, neither aluminum nor boron segregates normally. Aluminum has an anomalously large effective distribution coefficient ($k_{eq}=0.1$, $k_{eff} = 0.7$). This large effective distribution coefficient for aluminum is always observed in high-purity germanium grown from an SiO_2 crucible. This effect has been attributed to either an interaction at the melt/SiO_2 crucible interface [9,15,18] or to complexing of the aluminum in the melt [16]. For boron, an initial effective distribution coefficient of 3.5 is reasonable, but the long boron tail extending into the crystal indicates a continuing source of boron. The most important residual acceptors in high-purity germanium therefore appear to be reactive during growth.

PTIS of high-purity silicon has also been reported [6,19,20]. However, since boron has been the only residual acceptor and phosphorus the only residual donor observed, routine identification of the residual impurities in high-purity silicon is usually not necessary. Photothermal ionization spectra of high-purity silicon samples are shown in Fig. 5; note that the shallow levels are at higher energies in silicon than in germanium. This is due to larger effective masses and a smaller dielectric constant for silicon.

In both Fig. 3 and Fig. 5 the compensating impurities are neutralized by the minority carriers generated by illuminating the sample with band-edge light. The detection of minority carrier centers in this way was first reported by Bykova, Lifshits, and Sidorov [21] on n-type germanium, and later by other workers on p-type germanium [15,19] and on n-type and p-type silicon [19]. The

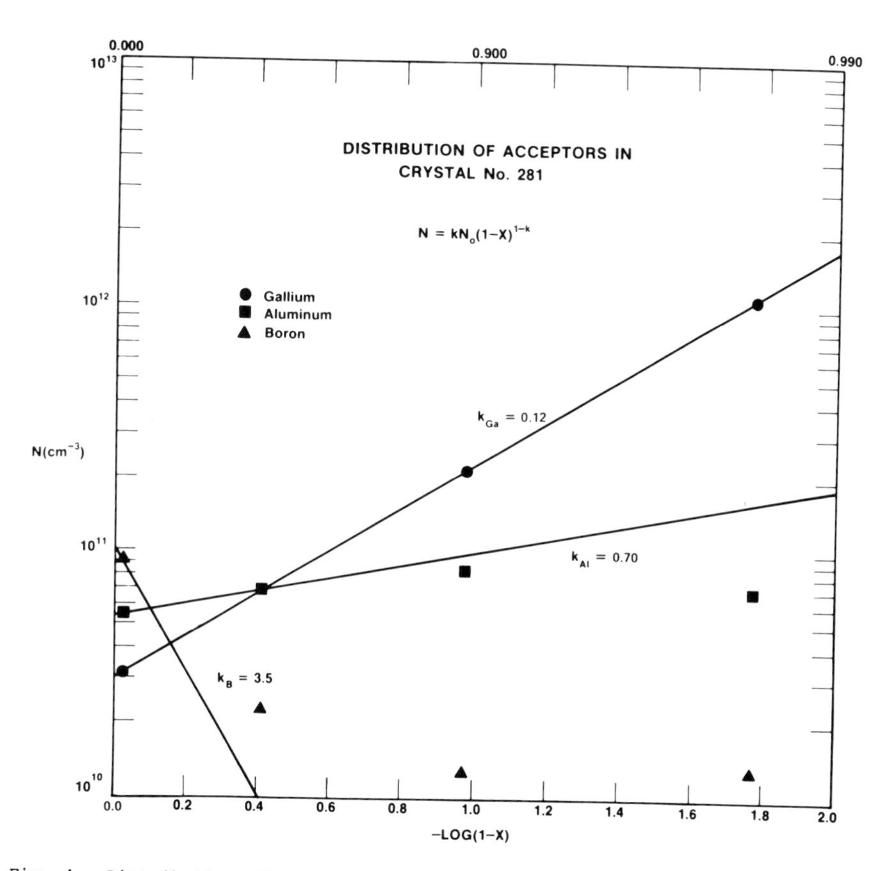

Fig. 4. Distribution of acceptors as determined by PTIS and Maxwell-Boltzmann statistics in a crystal doped with a known amount of gallium. The upper abscissa, X, is the position in the crystal normalized by mass and measured from the seed end. In the equation given for normal distribution, k is the distribution coefficient, and N_O is the original concentration of the impurity in the melt (Ref. 9).

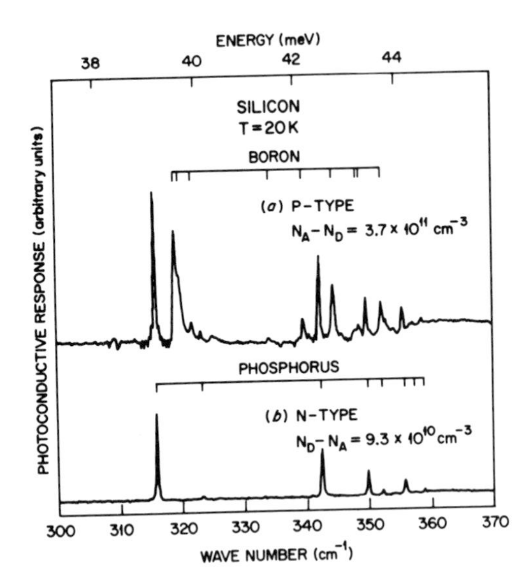

Fig. 5. Photothermal ionization spectra of silicon samples illuminated with band-edge light. In (a) a poi positive photoconductive response is obtained from the compensating phosphorus centers. In (b) no response from the compensating acceptor centers is observed (Ref. 20).

notable distinction between Figs. 3 and 5 is the different polarity of the compensating centers; phosphorus contributes negatively to the photoconductivity in the germanium sample and positively to the photoconductivity in the silicon sample. This reflects the diversity in the literature with both negative [9,15,19,21-23] and positive [6,23a,24] photoconductive responses from compensating centers reported. Generally, the negative photoconductivity has been attributed to rapid recombination of the minority carrier, thus removing a majority carrier. More recently, however, it has been shown that the ratio of the drift length ($\ell_D = V_D \tau_{relax}$) to the sample length (contact to contact) successfully predicts the polarity of the photoconductivity from the compensating centers in high-purity silicon and germanium [20]. If this ratio were smaller than one, the photoconductivity of the compensating centers was positive; if the ratio were larger than one, the photoconductivity was negative. The situation is illustrated in Fig. 6. In the first case shown, the drift lengths are smaller than the sample length, and the photothermally created minority carriers contribute positively to the photoconductivity. In the second case, the effective drift length of the majority carrier is several times the sample length. When a hole is swept to one p^+ contact, charge neutrality is maintained by injecting a hole from the opposite p^+ contact. However, when an electron is photothermally generated and swept to a contact charge neutrality is maintained by withholding a hole at the contact. The net effect of photothermally generating an electron is to remove a hole, thus decreasing the total carrier concentration and thereby also reducing the conductivity. With this model for data interpretation, PTIS can estimate the concentration of compensating impurities as well as identifying them.

PTIS has revealed several new shallow levels in germanium not associated with Group III or Group IV atoms [6-9]. Two of these were first observed without benefit of PTIS by Hall [15,25] in samples quenched to room temperature from 400°C. In these samples shallow acceptors were generated. Annealing slightly above room temperature removed these acceptors, and a donor appeared

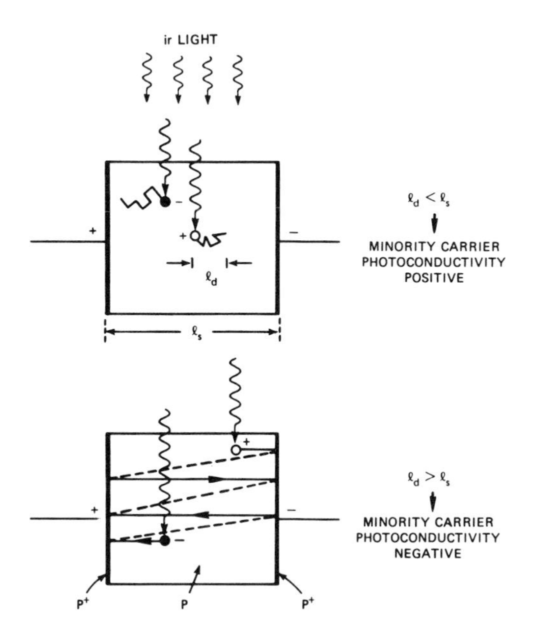

Fig. 6. Schematic representation of situations producing positive and negative photoconductivity by the photothermal generation of minority carriers. The semiconductor is p-type with p^+ contacts.

which could be annealed out at 150°C. Hall proposed that these "fast" centers were related to oxygen. It was subsequently shown that these defects occurred only in material containing hydrogen [26,27] and that, in fact, there was a hydrogen isotope shift in the ground state of the acceptor [27]. A scheme has been proposed by Haller, Joos, and Falicov identifying the acceptor as a hydrogen-silicon complex A(H,Si) [28] and the donor as an oxygen-hydrogen complex D(H,O) [29]. Other shallow-level complexes such as A(H,C) [28] and D(Li,O) [30] have also been seen by PTIS. Specific models for these complexes invoke a light atom (H or Li) tunneling around a heavier atom (Si, C, or O), resulting in additional splitting of the 1S-like ground states.

These complexes generally do not limit the net electrically active impurity content of high-purity germanium, since many of the complexes reported are only observed after anneals or quenches not required by detector processing. It is tempting to speculate whether analogous complexes can be generated in high-purity silicon.

DEEP LEVELS

Defect levels deeper than the hydrogenic levels are defined as deep levels. In addition to influencing depletion voltage and the spatial dependence of the electric field in the same manner as shallow-level defects, deep-level defects can also affect detector performance through the kinetics of carrier capture and emission. For example, states near the center of the energy gap are most effective in generating leakage current by the Shockley-Read-Hall mechanism. However, their most detrimental aspect is their ability to trap carriers. Good resolution in semiconductor nuclear radiation detectors requires that virtually all the free carriers created by the ionizing radiation be swept to a contact.

Therefore, the effective drift lengths must be much longer than the depletion depth: ~10 cm for silicon and ~100 cm for germanium. Thus, detector resolution is extremely sensitive to deep levels which can trap carriers for times comparable to or longer than the time constant of the associated detector electronics. The minority carrier lifetime, widely used in the semiconductor industry as an indicator of deep levels, is not generally sensitive or specific enough to be a useful parameter for detector-grade material. A good minority carrier lifetime does not necessarily guarantee good charge collection, and the lifetime limiting centers are not identified. PTIS of deep levels is too insensitive to be of practical use since the optical absorption cross-section of deep levels is much smaller than for shallow levels. Also, many deep states seem to have mostly nonradiative mechanisms for capturing excess carriers, and their detection by photoluminescence is ineffective [31].

Hall effect vs temperature

Until the mid 1970's the most widely used technique for detecting and identifying deep levels was measurement of the Hall coefficient as a function of temperature. A deep level present in dislocation-free germanium was shown in Fig. 1; this level at 0.072 eV above the valence band has been attributed by Haller et al. to a divacancy-hydrogen complex [32] and may be an important residual deep-level defect even in dislocated germanium. Hall coefficient measurements have also been useful in detecting deep levels in high-purity silicon. Levels at 0.26 eV and 0.40 eV above the valence band have been found in as-grown material [33]. The deeper level is probably gold.

The energy level deduced from Hall effect vs temperature measurements depends on the assumed value of degeneracy of the defect ground state, and the deduced concentration depends on the assumed value of θ. While there is some uncertainty in both, neither represents a significant impediment in characterizing high-purity semiconductors. The technique is limited to the top part of the energy gap in n-type material and to the bottom part of the energy gap in p-type material. In addition, intrinsic conduction may mask the thermal ionization of levels deeper than a certain value. For example, only the bottom third of the band gap is accessible to this technique in p-type high-purity germanium. While this technique is still useful, capacitive transient techniques, particularly deep level transient spectroscopy (DLTS), have demonstrated general superiority in sensitivity, energy range, and ability to resolve closely spaced levels.

Deep level transient spectroscopy

The origin of the capacitance transient in DLTS is illustrated in Fig. 7. The capacitance during and after a bias pulse is shown on the left in Fig. 7, and the charge state of the shallow (o) and deep (□) acceptors at various times is shown on the right. The time constant τ for thermal emission from the deep acceptors is a strong function of temperature and of the binding energy E_T of the hole. Using the principle of detailed balance, τ is given by

$$\tau = C_o T^{-2} \exp(E_T/kT) \quad , \qquad (4)$$

where C_o is independent of temperature if the trapping cross section σ_T is independent of temperature.

The magnitude of the capacitance transient is proportional to the concentration of the level if the deep level concentration is much smaller than the shallow level concentration. The capacitive transient can be indirectly measured displayed by various experimental schemes, the most common of which is the rate window technique [2]. By varying the rate window (or the characteristic measurement time of some other experimental scheme) E_T and σ_T can be obtained [35].

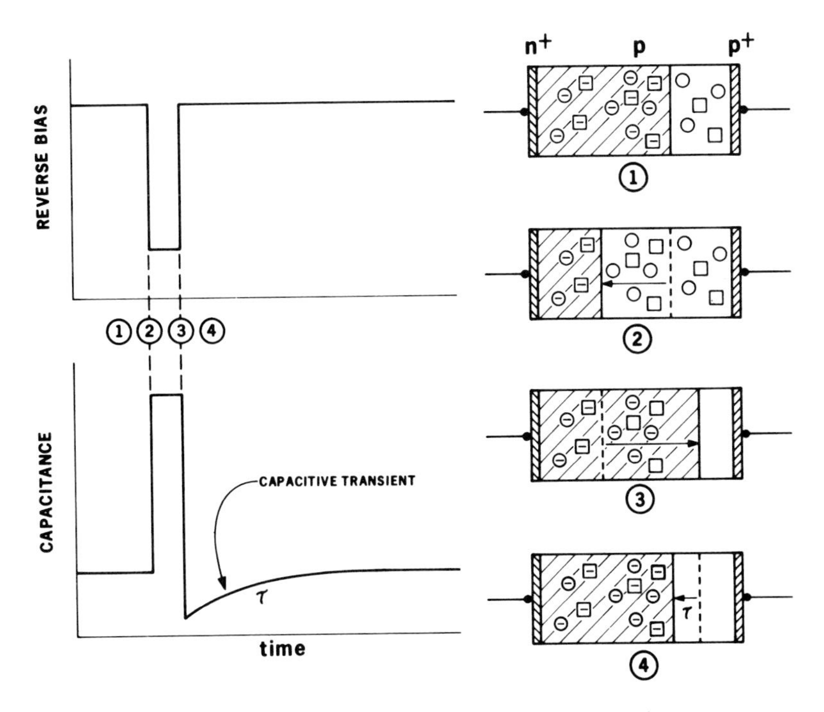

Fig. 7. Time dependence of bias and capacitance of a partially depleted p-n or p-i-n semiconductor junction in a deep level transient spectrometer. The changes of the depletion layer and the charge state of shallow-level (circles) and deep-level (squares) acceptors are shown on the right. Shallow acceptors follow the bias changes quickly, while the deep traps ionize more slowly with a time constant (Ref. 34).

Several hole traps in germanium have been investigated by DLTS [34,36]. The DLTS spectra shown in Fig. 8 and Fig. 9 were taken by Haller, Li, Hubbard, and Hansen [34] using the correlation technique [37]. Figure 8 shows the DLTS spectrum from a dislocation-free sample grown under a hydrogen ambient. Peak 7 is due to the vacancy-related level at E_V + 0.072 eV, also indicated in Fig. 1. This level was first observed by Hall [38] in Hall effect vs temperature measurements, and renders dislocation-free germanium grown under hydrogen unsuitable for radiation detectors due to poor charge collection. It may also be an important defect in lightly dislocated (10^2-10^3 cm^{-2}) germanium.

The DLTS spectra of various copper-doped samples are shown in Fig. 9. Peak 4 is the doubly ionized acceptor level (Cu^{--}), and peaks 1 and 3 are due to copper-hydrogen complexes. The singly ionized acceptor level (Cu^{-}) is offscale on the low temperature side. The copper-hydrogen complex is one of many instances of the reactivity of hydrogen in high-purity germanium. A relationship between total copper content (including complexed copper) as determined by DLTS and the resolution (FWHM for 1.33 MeV γ-rays) of p-type coaxial detectors has been reported by Simoen and coworkers [3] and is shown in Fig. 10. The

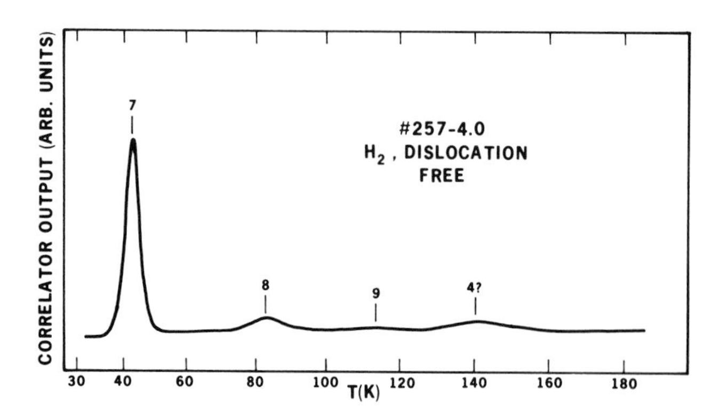

Fig. 8. Deep level transient spectra of dislocation-free high-purity germanium. Peak 7 is due to a vacancy-related complex. Peaks 8 and 9 have not been identified (Ref. 34).

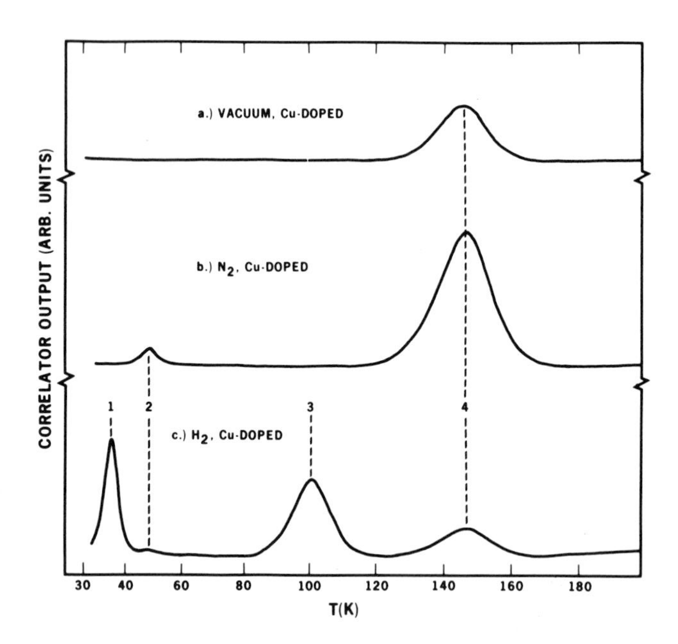

Fig. 9. Deep level transient spectra of three high-purity germanium doides which were intentionally doped with copper (Ref. 34).

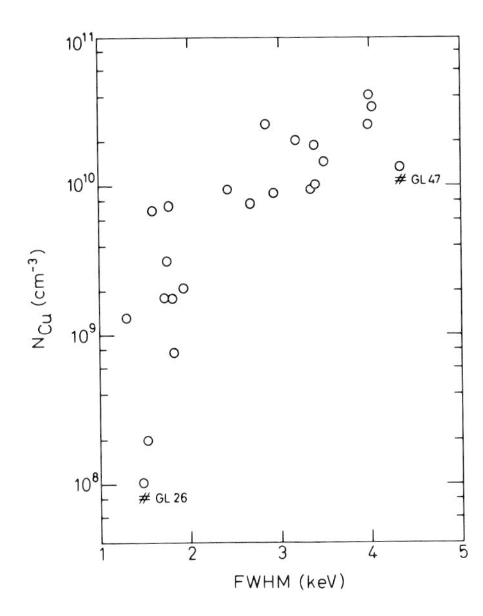

Fig. 10. Correlation of germanium coaxial detector resolution with concentration of active copper centers, as derived from DLTS. N_{Cu} is the sum of the concentrations of substitutional and complexed copper (Ref. 3).

correlation is apparent: all detectors with poor resolution had a high copper content. For detectors with resolutions less than 2.0 keV, DLTS samples were taken from crystal slices adjacent to the detector. For those detectors with resolutions greater than 2.0 keV, the DLTS samples were cut directly from the detector. This alert to the significance of copper contamination reemphasizes a long time concern in germanium detector technology [39].

This relative abundance of DLTS data on hole traps in germanium contrasts with the lack of published data available on electron traps. More attention is expected to be paid to electron traps with the increased use of the n-type coaxial detector geometry [40]. These devices use a contact polarity (the reverse of the normal p-type coaxial configuration) which emphasizes electron traversal.

Similarly, there are no DLTS data published on high-purity silicon for nuclear radiation detectors. However, silicon detectors have been produced from high-resistivity, neutron-transmutation-doped silicon [42], and the annealing of electron traps induced in transmutation doped silicon (1.2×10^{14} phosphorus atoms/cm^3) has been studied by DLTS [41]. Charge collection distances are usually shorter in silicon than in germanium detectors, and, therefore, resolution is less sensitive to carrier trapping. Thus, perhaps the need to identify deep levels in silicon has not been as urgent as in the case of germanium.

Transition and noble metals

The transition metals and noble metals form deep levels in silicon and germanium and can diffuse relatively quickly via interstitial sites. They may be incorporated during growth or introduced during subsequent thermal processing for contact formation. There is a possibility of contamination by those elements with significant solubilities at the processing temperature. One source of these contaminants may be trace metallic impurities in an etchant which can electrochemically plate onto the semiconductor during etching. On the other hand, the high diffusivity and solubility of these impurities also makes it possible to getter them by making the surface a sink rather than a source. This may be done by physical or chemical methods (surface damage, gas phase gettering, liquid phase gettering, etc.).

Copper is a fast-diffusing impurity with a relatively high solubility in germanium at low temperatures. Thus, in addition to copper being incorporated into the crystal during growth, it is the most likely contaminant to be introduced into the bulk during thermal processing. Copper solubility at 400°C, as measured by DLTS using the singly ionized level at $E_v + 0.044$ eV, has been reported to be strongly dependent on growth conditions and dislocation density [7]. For moderately dislocated ($\sim 10^3$ cm^{-2}) high-purity germanium grown under hydrogen in an SiO_2 crucible, the solubility was $[Cu]_o = 8.2 \times 10^{11}$ cm^{-3}, while in dislocation-free germanium the solubility was $[Cu]_o = 4.8 \times 10^{12}$. The higher solubility of copper in dislocation-free germanium had been reported earlier [36].

The high solubility and fast diffusivity of copper in germanium makes the gettering of copper by an external phase an attractive possibility. Attention seems to have been focused on liquid-metal gettering near 400°C, where lead-bismuth eutectic [26] and gold [7] have been used. Effective distribution coefficients on the order of 10^{-10} are expected for copper between the semiconductor and the liquid-metal phase. The copper-germanium system is not simple; in addition to precipitating at low temperature into a copper-rich phase, copper may complex with hydrogen, oxygen, or lithium to form electrically active levels. Thus the performance of a detector would be affected not only by total copper content, but also by thermal history as well. While the deleterious effects of copper are well substantiated, a complete and realistic model for its low temperature (< 400°C) kinetics and equilibria in high-purity germanium is not available.

Data on the solubility of fast diffusing defects, and on the identification of thermal defects have been obtained by DLTS in silicon. The solubility of some transition and noble metals in dislocation-free silicon, obtained by Graff and Pieper [43] using DLTS, is shown in Fig. 11. The retrograde solubility of copper and nickel is probably an artifact of precipitation on quenching. Despite the greater solubility of copper and nickel in silicon, the most common impurity introduced by heating silicon is iron [44,45] (most reports are on material quenched from near 1000°C). This is probably not important to surface barrier technology but may be an important limitation to the diffused-junction detector technology. Iron is an important impurity both in the grown material [44] and in finished integrated devices [46].

Generation current

Deep levels can also act as generation centers for leakage current. This generation current i_g is given by [47]

$$i_g = n_i A w / 2\tau_o \quad , \qquad (5)$$

where n_i is the intrinsic carrier concentration; A, the device area; w, the depletion depth; and τ_o, the minority carrier lifetime. For silicon at room temperature with w = 1 mm and $\tau_o = 10^{-3}$ sec (a representative value for detector-grade silicon), Eq. (5) implies $i_G/A = 0.13\ \mu A/cm^2$. The emission over a good Si/Au surface barrier (0.8 eV) is 0.1-0.5 $\mu A/cm^2$ [48]. Thus for thicker silicon devices, bulk generation can be a significant contribution to the overall leakage current [49].

For germanium detectors operated near 77 K, n_i is small enough that bulk generation is usually not a problem. Using a buried juntion "monode" device,

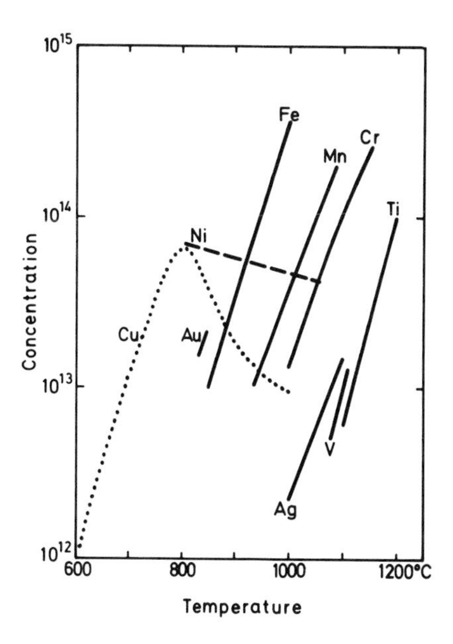

Fig. 11. Defect state concentrations in silicon determined by DLTS after diffusion, saturation and quenching of transition and noble metals (Ref. 43). This figure was originally presented at the Spring 1981 meeting of the Electrochemical Society, Inc. held

Hall [50] has shown that in depleted high-purity germanium the generation current exceeds the current produced by background ionizing radiation only above 120 K. In conventional germanium detectors, surface currents are the most troublesome contribution to the leakage current.

NEUTRAL IMPURITIES

Several neutral species are usually present in higher concentrations than the electrically active species: oxygen and carbon in high-purity silicon, and oxygen, carbon, hydrogen, and silicon in high-purity germanium. These neutral impurities can affect device fabrication and performance by their reactivity. They may precipitate, act as nucleation sites for precipitation, or complex with other point defects. For example, the complexing of lithium with immobile defects such as oxygen reduces both the lithium diffusion coefficient and drift mobility [51]. It should also be noted that these neutral impurities may also be reactive during growth; electrically active impurities may form volative hydrides, hydroxides, or oxides.

Table I presents estimates of the concentrations of the residual neutral impurities in high-purity silicon [52] and germanium. The values are for silicon as grown by the float-zone process under vacuum or argon and for germanium by the Czochralski method under hydrogen from an SiO_2 crucible. Techniques for the detection of neutral impurities are less sensitive than those for electrically active defects. Standard analytical techniques such as neutron activation analysis and spark source mass spectroscopy are usually not sensitive enough to be useful in the analysis of these high-purity semiconductors. Therefore, specific analytical techniques were developed in several cases.

TABLE I
Neutral impurities

Defect \ Host	Si^{a}	Ge^{b}
C	$< 10^{16}$ cm^{-3}	$\lesssim 10^{12} - 10^{13}$ cm^{-3}
O	$\lesssim 10^{15}$ cm^{-3}	$10^{13} - 10^{14}$ cm^{-3}
H	-	$\sim 10^{15}$ cm^{-3}
Si	-	$\lesssim 10^{15}$ cm^{-3}

a)Typical for float-zone silicon.
b)SiO_2 crucible, H_2 ambient.

Neutral defects in silicon

High-purity silicon for either surface barrier detectors or lithium-drifted Si(Li) detectors is float-zoned under either vacuum or argon. The residual amounts of oxygen and carbon can be below the detection limits for optical absorpton at 77 K and for neutron activation analysis: $2x10^{15}$ for oxygen and $3x10^{16}$ for carbon [53]. In less pure electronic-grade silicon much effort has been expended in studying the generation of defect clusters or "swirls" [54] and the effects of the concentration of oxygen and carbon on the generation of these defects. Little has been reported on the effects of "swirls" in high-purity silicon for nuclear radiation detectors. While one would expect fewer such microdefects, device resolution might be very sensitive to them.

Oxygen may react to some extent with the Group III acceptors. In boron-doped, float-zoned silicon (10 Ω-cm) it has been suggested that the boron forms compensating complexes with residual oxygen [53]. The complexing of aluminum with oxygen to form deep levels has also been reported [55].

Hydrogen and carbon in germanium: the beta decay technique

More is known about the quantity of neutral impurities and their various effects in high-purity germanium than in high-purity silicon. Specialized techniques have been developed to determine the concentrations of hydrogen, carbon, and oxygen in high-purity germanium. For example, the hydrogen and carbon contents of high-purity germanium have been measured by incorporating radioactive isotopes of these elements during growth, fabricating detectors from this material, and observing the count rate from the beta particles emitted during nuclear decay.

The role of hydrogen in forming electrically active defect complexes with copper, oxygen, silicon, and carbon has already been mentioned, and has been emphasized by Haller [56]. Pearton [57], using DLTS, has demonstrated the effectiveness of hydrogen from a plasma source in removing deep copper levels from the surface region (~80 μm) of a germanium crystal held at 300°C. Hall and Soltys [26] have measured the diffusion coefficient of hydrogen in germanium at 400°C by a technique using the shallow-level hydrogen complexes; the value obtained was two orders of magnitude lower than that extrapolated from higher temperature [58]. This immobility indicates that the hydrogen was largely bound in some neutral complex such as a hydrogen molecule.

In dislocation-free, hydrogen-grown germanium smooth pits are observed after a chemical etch intended to reveal dislocations [59]. Their concentration has been determined by stripping and counting to be roughly $1x10^{7}$ cm^{3} [5]. Presumably in dislocated germanium many more smaller, but similar, defects decorate the dislocations. It was suggested on the basis of the high temperature (900°C) stability of these defects that they are not interstitial-type precipitates but are rather bubbles of hydrogen [5].

Hansen, Haller and Luke [5] have measured the concentration and distribution of hydrogen in high-purity germanium. They constructed a Czochralski furnace in which the hydrogen was recirculated and purified in a closed system and in which tritium (a beta emitter) could be generated and recovered by means of heating and cooling a uranium pellet. Hydrogen content was deduced from the tritium/hydrogen ratio in the furnace during growth and the count rate in detectors fabicated from this material. The distribution of hydrogen in a crystal is shown in Fig. 12. The uneven distribution was attributed to the out-diffusion of hydrogen while the crystal was cooling. From the temperature gradient and growth rate of the crystal, a diffusion length for hydrogen of ~1 cm at the shoulder was estimated, based on the diffusion coefficient data of Frank and Thomas [58]. This is consistent with the higher hydrogen content at the radial center and tail of the crystal. It is noteworthy that the hydrogen concentrations indicated in Fig. 12 are greater than the solubility at the melting point also determined by Frank and Thomas (4×10^{14} cm^{-3}) [58]; this discrepancy was attributed to a supersaturation of hydrogen at the melt/solid interface during growth.

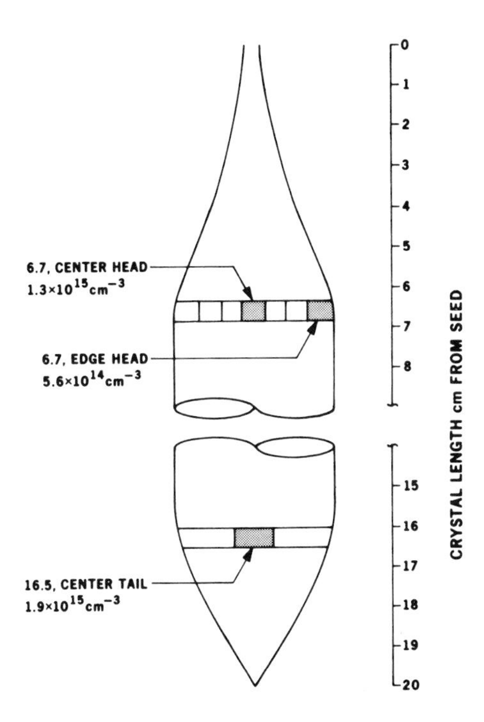

Fig. 12. Location and hydrogen concentrations of samples taken from a tritium-grown crystal (Ref. 5).

The carbon content of high-purity germanium has been measured by Haller, Hansen, Luke, McMurray, and Jarrett [4] using a 14C-spiked pyrolytic graphite coating on the quartz crucible. Carbon contents were subsequently deduced from the count rate of the emitted beta particles in fabricated detectors. An average value for the total carbon concentration determined was roughly 2×10^{14} cm^{-3}. A second generation crystal regrown under standard conditions for high-purity germanium (H_2 ambient, SiO_2 crucible) contained $\sim10^{13}$ cm^{-3} carbon. Shallow-level acceptor complexes containing carbon have only been observed in rapidly quenched germanium grown from graphite crucibles [28].

Oxygen in germanium: The lithium precipitation technique

The oxygen content of high-purity germanium is measured by the lithium precipitation technique. The germanium sample is saturated with lithium near 425°C, quenched to room temperature, and the carrier concentration monitored as the lithium precipitates [60]. The lithium also reacts with oxygen to form a donor complex,

$$\underline{Li}^+ + \underline{O} = \underline{LiO}^+ \quad , \tag{6}$$

which at room temperature proceeds to near exhaustion of the dissolved oxygen. Terms are underscored to indicate that they are in solid solution. If the donor concentration is dominated by $\underline{LiO}^+$, the oxygen content is given by the final electron concentration in the precipitated sample. This technique was firmly established by Fox [61] who showed a linear relationship beween oxygen content in the range 10^{13}–10^{14} cm^{-3} as measured by lithium precipitation and by the infrared lattice absorption of oxygen at 4 K. The validity of the technique was later widened to 10^{12}–10^{15} cm^3 when it was demonstrated that over this range the oxygen content as measured by lithium precipitation agreed with the oxygen content predicted by thermodynamic equilibrium with the P_{H_2O}/P_{H_2} ratio during growth [62]. This relationship is shown in Fig. 13, where the oxygen content of several samples, as determined by the lithium precipitation

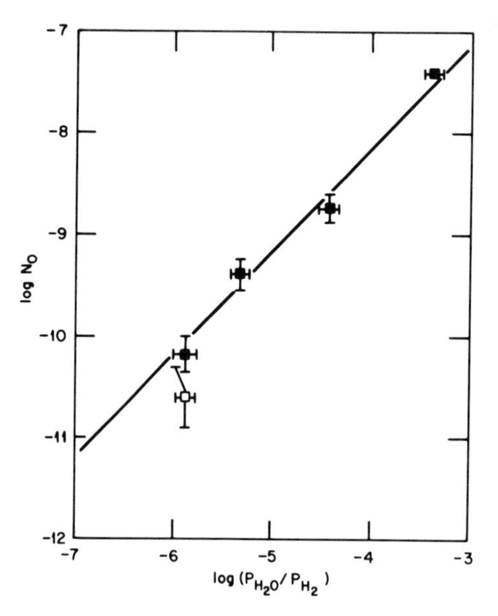

Fig. 13. Atom fraction of oxygen in germanium crystals for different P_{H_2O}/P_{H_2}. The unfilled data point was obtained without taking precautions against copper contamination during measurement of the oxygen content by the lithium precipitation technique. The line is a thermodynamic calculation, not a fit to the data points. The figure is reprinted by permission of the publisher, The Electrochemical Society, Inc. (Ref. 62).

technique, is plotted as a function of the ratio of the partial pressure of water vapor to the partial pressure of hydrogen during growth of the samples. The line is a thermodynamic calculation independently obtained; the excellent agreement between experiment and calculation indicates that the lithium precipitation technique is generally valid over range [O] = 10^{12}-10^{15} cm^{-3}.

The lithium precipitation technique has natural limitations. For the technique to be applicable it is necesary that the oxygen content be significantly greater than the intrinsic solubility of lithium $[\underline{Li}^+]_o$ and the intrinsic carrier concentration n_i. In germanium at room temperature $[\underline{Li}^+]_o \simeq n_i \simeq 2x10^{13}$ cm^{-3}. The lithium precipitation technique has been extended to lower oxygen contents ($\sim 10^{12}$ cm^{-3}) by monitoring precipitation at -22°C [62].

The lithium precipitation technique at high oxygen contents is limited by a more subtle effect. Since this limit has not been previously pointed out, a full account will be presented here. It is well extablished that the position of the Fermi level influences the solubility of lithium (and more generaly all donors and acceptors) in semiconductors [63,64]. It is more energetically favorable for a donor to dissolve in p-type material by the additional energy the electron loses in going from the conduction band to the valence band. Thus the solubility of lithium is increased in p-type germanium and decreased in n-type germanium. Since $\underline{LiO}^+$ is a donor, $[\underline{Li}^+]_o$ is suppressed by $[\underline{LiO}^+]$. By inspection of Reaction (6), it can be seen that this effect will increase the concentration of uncomplexed oxygen.

In order to determine when the concentration of uncomplexed oxygen becomes significant, the equilibrium constants of the relevant reactions have to be considered:

$$Li(ext.\ phase) = \underline{Li}^o \ , \quad K_o = \frac{[\underline{Li}^o]}{[Li(ext.\ phase)]} \ ; \tag{7}$$

$$\underline{Li}^o = \underline{Li}^+ + e^- \ , \quad K_D^{Li} = \frac{[\underline{Li}^+]n}{[\underline{Li}^o]} \ ; \tag{8}$$

$$\underline{Li}^+ + \underline{O} = \underline{LiO}^+ \ , \quad K_{LiO} = \frac{[\underline{LiO}^+]}{[\underline{Li}^+]\ [\underline{O}]} \ ; \tag{9}$$

$$np = n_i^2 \ ; \tag{10}$$

as well as the charge neutrality condition:

$$[\underline{Li}^+] + [\underline{LiO}^+] + p = n \ . \tag{11}$$

Superscripts indicate charge state, and the external lithium phase in Reaction 7 is the precipitated phase. Solving Eqs. (7)-(11) for $[LiO^+]$ we obtain

$$[\underline{LiO}^+] = \frac{K_{LiO}K^*}{(K^* + K_{LiO}K^*[\underline{O}] + N_i^2)^{1/2}} [O] \ , \tag{12}$$

where

$$K^* = K_D^{Li} K_o[Li(ext.\ phase)] = [\underline{Li}^+]\ n \ . \tag{13}$$

According to Reiss and Fuller [65]

$$K^* = \frac{[\underline{Li}^+]_o^2}{2} + \left\{ \frac{[\underline{Li}^+]_o^4}{4} + n_i^2 \, [\underline{Li}^+]_o^2 \right\}^{1/2} \tag{14}$$

where $[Li^+]_o$ is now restricted to mean the lithium solubility in intrinsic material. Evaluating K^* using $n_i = 1.92 \times 10^{13}$ cm^{-3} and $[Li^+] = 2.23 \times 10^{13}$ cm^{-3} [62] we obtain $K^* = 7.44 \times 10^{26}$ cm^{-6}. Since we are interested in the case $[\underline{LiO}^+]/[\underline{Li}^+] >> 1$, Eq. (9) then implies that in this limit $K_{LiO}\,[\underline{O}] >> 1$. Thus the middle term under the radical in the numerator on the right hand side of Eq. (12) is much larger than the two other terms. Ignoring the other two terms we simplify Eq. (11) to

$$[\underline{O}] = [\underline{LiO}]^2/K_{LiO}K^* \quad . \tag{15}$$

Thus the uncomplexed oxygen content is proportional to the square of the complexed oxygen content and therefore must dominate at high enough total oxygen concentrations.

In order to ascertain when this effect is significant, K_{LiO} must be known. Fox [61] has suggested the value $k_{LiO} = 5 \times 10^{-13}$. From Eq. (15) this would imply $[\underline{LiO}^+]/[\underline{O}] = 1$ at $[\underline{LiO}^+] = 3.7 \times 10^{14}$ cm^{-3}. The good agreement shown in Fig. 13 at $[LiO^+] = 10^{15}$ cm^{-3} suggests a larger value for K_{LiO} would be more appropriate ($K_{LiO} > 2 \times 10^{-12}$). For the range of oxygen concentrations normally found in high-purity germanium (10^{13}-10^{14} cm^{-3}) a correction to the lithium precipitation technique to account for the uncomplexed oxygen is not required. This lower limit on K_{LiO} also confirms the proposition that at dilute lithium concentrations the equilibrium in Reaction (6) is strongly in favor of high $[\underline{LiO}^+]/[\underline{Li}^+]$ ratios in high-purity germanium grown from SiO_2 crucibles ($[\underline{LiO}^+]\ [\underline{Li}^+] > 80$ for $[O] = 4 \times 10^{13}$ cm^{-3}).

The complexing of lithium with oxygen was a major problem in compensating germanium by lithium drifting for Ge(Li) detectors [51], since the drift mobility of the lithium was reduced by the fraction of lithium that was complexed. In addition to reactions in the solid state, the reactivity of oxygen with impurities in the melt has also received deserved attention [16,62,66]. Understanding such reactions would aid process control in high-purity germanium growth. The solubility of oxygen in liquid germanium has been measured by direct phase analysis [67] and electrochemically [68]. Both techniques obtain a maximum oxygen content at the melting of $N_{O(l)} = 7.1 \times 10^{-4}$ (in atom fraction). Combining this with the extrapolated maximum solubility in the solid phase of $N_{O(s)} = 1.43 \times 10^{-4}$ [62,69,70], an equilibrium distribution coefficient of k=0.20 for oxygen between the liquid and solid phases is obtained. Negligible change of the activity coefficient of oxygen in liquid germanium over the solubility range is assumed. Rough agreement with the 0.11 value for the effective distribution coefficient measured indirectly by Edwards [70] using an aluminum doping technique should be noted. For comparison an equilibrium distribution coefficient of 1.25 for oxygen in silicon has been determined [71].

This distribution coefficient obtained for oxygen in germanium and the relationship between $[\underline{O}(s)]$ and P_{H_2O}/P_{H_2} obtained in Ref. 62 imply a ratio of 1.9 between the absolute concentration of water in the gas phase ($P_{H_2} = 1$ atm) and the concentration of oxygen in the liquid germanium phase, both at 1200 K. This relationship determines the parameters for effective buffering of oxygen by the gas phase.

Silicon in germanium

Silicon is expected in high-purity germanium due to the difficulty in removing this element and the generation of silicon from the dissoluton of the SiO_2 crucible. However, Hall could not detect silicon in his high-purity germanium with spark source mass spectroscopy [15], putting an upper limit on the silicon content of about 10^{15} cm^{-3}. Neutron activation analysis for silicon would not be more sensitive.

In addition to forming shallow-level complexes (such as A(H,Si)), silicon is believed to be the first impurity to precipitate into a solid oxide phase if the chemical potential of oxygen during crystal growth is increased. Dislocation etches on high-purity germanium grown with $P_{H_2O}/P_{H_2} > 10^{-3.5}$ reveal an abundance of shallow etch pit defects [66] and charge collection is very poor in such material. While the defects responsible have not been directly identified, the strong presumption is that they are silica or silicate [66,72] precipitates generated in the melt and included during growth.

EXTENDED DEFECTS

In this section various experimental results and possible effects of extended defects will be presented in order to provide an overview to a subject where, due to its complexity, basic understanding is incomplete. The most important extended defects are dislocations and particulates. There is no widely accepted microscopic model for the local states or energy bands associated with either defect. However, the effects of these defects on detector performance have been quantified in some cases.

Dislocations

Dislocations are a common crystallographic defect and may affect the performance of nuclear radiation detectors in several ways: they contribute directly to net space charge, trap carriers generated by ionizing radiation, and interact with point defects. These complicating features are avoided in the case of silicon by using only dislocation-free silicon for both Si(Li) and surface barrier detectors. However, as has already been mentioned, excessive charge trapping is observed in detectors made from dislocation-free germanium grown under hydrogen.

Theoretical models or experimental data are not available to predict the charge state or trapping cross-sections of dislocations in the space charge region of a reverse-biased junction [73]. In fact, the origin of the electrical activity of dislocations in semiconductors is in doubt. It was proposed by Shockley [74] and further developed by Read [75] that dislocations in the diamond structure created dangling bonds possessing localized electronic states which were broadened into a band structure. This theory was generally accepted until it was revealed by the weak-beam method in tranmission electron microscopy that dislocations in silicon and germanium are usually dissociated into partial dislocations over most of their length [76,77]. Theoretical calculations indicate that the dangling bonds may vanish in this reconstruction [78,79]. It has been suggested that the electrical activity of dislocations can be due to residual dangling bonds [80], banding in the reconstructed core and kink sites [81,82], or controlled by a cloud of point defects surrounding the core [83]. In the absence of a generally accepted theoretical viewpoint, experimental results have to be examined individually and extrapolated with caution. However, some results on heavily dislocated electronic-grade semiconductors may provide general insight into the effects of dislocations on nuclear radiation detectors. Ourmazd and Booker [84] have examined the correspondence between the core structure of dislocations and their electronic properties by probing dislocations with both weak-beam tramsmission electron microscopy and the electron-beam induced-current mode of the Scanning electron microscope.

They found that dissociation of the <110> a/2 edge dislocation into partial dislocations significantly increased recombination efficiency. This is direct proof that the core structure of a dislocation strongly affects the electrical behavior.

Schroter, Schiebe and Schoen [80] have emphasized the strong effect of the temperature at which the dislocations are introduced. In germanium there is a much stronger electrical activity associated with dislocations introduced below 450°C. Likewise whether the dislocation is screw- or edge-type affects electrical properties [85]. Thus we expect that thermal factors during growth, growth ambient, as well as the presence of other defects, may affect the electrical properties of dislocations.

Glasow and Haller [86] analyzed planar detectors from a germanium crystal grown on the <100> axis and found rapid resolution degradation as the dislocation etch pit density (EPD) increased beyond 10^4 cm^{-2}. Later Hubbard, Haller, and Hansen [87] reported that detectors made from germanium crystals grown on the <113> axis were not degraded at this EPD. Hall effect vs temperature and DLTS measurements identified two dislocation-related bands in <100>-grown germanium that were shifted closer to the valence band in <113>-grown germanium [88], which could account for the lesser effect of dislocations in <113>-grown germanium in degrading charge collection. It was also found that the position and breadth of these dislocation bands depended on whether the germanium was grown under H_2 or N_2 [88]. The mechanism whereby growth direction or ambient influences the electrical properties of dislocations in germanium could not be determined. However, the net line charge on a dislocation could shift the energy levels of the single-electron states associated with the dislocation [89]. It has also been observed in DLTS of deformed silicon that the capture process is proportional to the logarithm of time [90]; this is characteristic of large defects with many charge states surrounded by a space charge region. Thus the techniques for measuring isolated deep levels require care when applied to the deep levels associated with dislocations.

It is interesting that in conventional silicon diodes and transistors direct effects of dislocations are hard to find; they appear to affect leakage current only when decorated with precipitates [91]. The most important aspect of dislocations may not be their own intrinsic electrical activity but their chemical reactivity. Dislocations are widely believed to maintain the equilibrium between vacancies and interstitial during the cool-down of the crystal after growth. It is the supersaturation of vacancies that is responsible for the vacancy-related hole traps in dislocation-free germanium. Dislocations are also effective nucleation sites for precipitation: hydrogen [5] and copper [92] in germanium and iron in silicon [93]. High leakage current in conventional diodes and transistors has been correlated with dislocations [91,94,95] and stacking faults [93,96,97] intersecting the junction. However, only Bull et al. [94] attribute this effect directly to the dislocations or stacking faults, while the others attribute the leakage current to precipitates decorating the dislocations or stacking faults. Holt [91] suggests that no effects of undecorated dislocations (EPD $< 10^4$ cm^{-2}) on conventional diode or transistor structures have been observed because these dislocations are usually neutralized by impurities.

Hall has observed that lineage affects leakage current in high-purity germanium diodes at 77 K [98]. Lineage with 5 μm spacing between dislocations has no noticeable effect on leakage current, while lineage with 1 spacing induces strong leakage current. The mechanism for this effect is unclear.

Precipitate and Inclusions

Dispersed particles of a second phase may be either included during growth from the liquid phase (inclusions) or nucleate and grow in the solid (precipitates). Such defects can cause either charge trapping or low breakdown voltages. Though the data on particulates is largely qualitative, experience generally has been that they can be much more harmful to detectors than dislocations. Silicon for both surface barriers and for Si(Li) detectors is usually specified by device fabricators to be free from "swirl" defects [54]. As discussed in the last section, shallow etch pits in germanium believed to be associated with oxide inclusions, have been correlated with poor charge collection [72,99]. The means by which these defects may affect device performance will be outlined briefly.

Precipitates and inclusions may increase leakage current by locally concentrating electric field lines. Metallic particulates concentrate electric field lines around points of sharp curvature, while dielectric particulates produce high field regions as the electric field lines bend around these particles. The magnitude of the field enhancement depends on the geometry of the particulate [100]. Reverse-bias leakage current due to transition and noble metal precipitates usually varies as a power law (I V^n, n= 4-7) [100,101]; above a certain onset voltage, this precipitation induced soft breakdown dominates the Shockley-Read-Hall generation current of the isolated defects [100]. In a nuclear radiation detector the effects of particulates on leakage current would be more important near regions of high fields created by the geometry of the contacts and by surface charges.

Random, high-amplitude, low-frequency noise pulses, "popcorn" noise, in conventional diode and transistor structures have been attributed to the shorting action of metallic precipitates [97,102]. Whether a mechanism involving precipitates is a common cause of "popcorn" noise in semiconductor nuclear radiation detectors is not known.

The origin of the electronic levels associated with particulates is usually not known; therefore, the effect of particulates on charge collection and other detector parameters cannot be predicted a priori. Electronic effects would likely depend on specific details of the structure and interface. However, these particulates are unlikely to act as passive voids in the detector. One would expect surface charges due to a variety of physical effects, including surface states and the difference electron affinity for the two materials. Recombination at dislocations has been successfully modeled as a recombination center having a large number of charge states and equating carrier trapping and emission [103]. Hole trapping by the disordered regions created by fast neutrons in depleted high-purity germanium detectors has also been treated by considering capture by, and emission from, a center with multiple charge states [104]. A similar model may also be applicable to carrier trapping by particulates.

SUMMARY

This review has been weighted toward germanium because greater and more diverse efforts have been expended in the characterization of high-purity germanium than of high-purity silicon. Several factors are relevant: historically, greater concern about the availability of germanium for radiation detectors than of silicon, the relative simplicity of the Czochralski growth of germanium compared to the float-zone growth of silicon, and the closer organizational coupling of crystal growth and detector making in germanium detector technology than in silicon. The nuclear detector community has been in the forefront of material research and development in germanium; in silicon it has not.

Some topics have not been included in this review. Foremost in this category is the compensation of semiconductors by lithium drifting. The lithium drifting of germanium is of strong historical interest, but the shift in research and development to high-purity germanium has been followed by the gradual replacement of Ge(Li) detectors by high-purity detectors. Thus, it is more forward looking to concentrate on the characterization of high-purity germanium. Lithium drifted silicon detectors continue to have unique advantages as x-ray detectors. However, published work on the effects of material parameters on the drifting process in silicon are scarce and recent work by Fong et al. [105] will be covered elsewhere in this symposium [106]. I have also chosen not to discuss the effects of bulk properties on diffusion and implantation.

A couple of general comments on the characterization of silicon and germanium for nuclear radiation detectors can be made. First of all, the basic understanding of different types of defects and their effects on detectors is uneven. Both shallow and deep levels are fundamentally sufficiently well understood, their concentrations can be measured, and their effects are at least semiquantitatively understood. However, the various roles played by neutral defects are neither fully known nor exploited. With respect to extended defects, we have only general ideas and some empirical data.

Secondly, a basic theme running through this review has been the extent of chemical reactions among defects even in these high-purity semiconductors: point defects forming complexes, point defects precipitating, and the role of point defects and dislocations as nucleation sites for precipitation. The underlying root of these phenomena is the existence of a temperature range of both high supersaturation and high mobility for a reacting defect. The importance of chemical reactions during purification and crystal growth has been emphasized by others at this symposium [52,107,108]. Some of these reactions in high-purity semiconductors have been beneficially exploited (either intentionally or unintentionally) and others are undesirable. In either case, an understanding of the mechanisms involved allows control of the material properties affected.

The characterization of defects in high-purity semiconductors can present unique challenges. However, the feedback provided by defect chaacterization has been, and will continue to be, crucial in optimizing the entire process from crystal growth to detector fabrication.

ACKNOWLEDGMENTS

I am grateful to numerous individuals for contributing useful discussions and advice: R. Baron, J. Emery, R. N. Hall, E. E. Haller, W. L. Hansen, S. A. Hyder, H. Kraner, R. Trammell, and J. Walton. Also E. E. Haller, W. L. Hansen, E. Simoen, and K. Graff are thanked for kind permission to use figures and figure captions. The author also acknowledges G. E. Jellison and L. H. Jenkins for their careful reading of the manuscript.

REFERENCES

1. This technique was first reported by T. M. Lifshits and F. Ya. Nad, Soviet Phys. Dokl. 10, 532 (1965), and the literature to 1977 was reviewed by Sh. M. Kogan and T. M. Lifshits, Phys. Status Solidi A 39, 11 (1977).

2. D. V. Lang, J. Appl. Phys. 45, 3022 (1974).

3. E. Simoen, P. Clauws, J. Broeckx, J. Vennik, M. van Sande, and L. De Laet, IEEE Trans. Nucl. Sci. NS-29, No. 1, 789 (1982).

4. E. E. Haller, W. L. Hansen, P. Luke, R. McMurray, and B. Jarrett, IEEE Trans. Nucl. Sci. NS-29, No. 1, 755 (1982).

5. W. L. Hansen, E. E. Haller, and P. N. Luke, IEEE Trans. Nucl. Sci. NS-29, No.1, 738 (1982).

6. E. E. Haller, Bull. Acad. Sci. USSR Phys. Ser. 42, 8 (1979).

7. E. E. Haller, W. L. Hansen, and F. S. Goulding, Adv. Phys. 30, 93 (1981).

8. P. Clauws, K. Vanden Steen, J. Broeckx, and W. Schoenmaekers in: Defects and Radiation Effects in Semiconductors-1978 (Institute of Physics, 1979), p. 218.

9. L. S. Darken, J. Appl. Phys. 53, 3754 (1982).

10. L. J. van der Pauw, Philips Res. Rep. 13, 1 (1958).

11. For a brief discussion see S. M. Sze in: Physics of Semiconductor Devices (John Wiley and Sons, New York 1981) p. 45; for a more extended treatment, see E. H. Putley, The Hall Effect and Related Phenomena (Butterworth, Inc., Washington, D.C. 1960).

12. For a recent review of photoluminescence see P. J. Dean, Prog. Crystal Growth Charact. 5, 89 (1982).

13. M. Tajima, T. Masui, T. Abe, and T. Iizuka in: Semiconductor Silicon - 1981 (The Electrochemical Society, New Jersey 1981) p. 72.

14. D. Itoh, I. Namba, and Y. Yatsurugi, this volume.

15. R. N. Hall, IEEE Trans. Nucl. Sci. 21, 260 (1974); E. M. Bykova, L. A. Goncharov, T. M. Lifshits, V. I. Sidorov, and R. N. Hall, Sov. Phys. Semicond. 9, 1223 (1976).

16. E. E. Haller and W. L. Hansen, Solid State Commun. 15, 687 (1974); E. E. Haller, W. L. Hansen, G. S. Hubbard, and F. S. Goulding, IEEE Trans. Nucl. Sci. 23, No. 1, 81 (1976).

17. For a low concentration ($\sim 10^7$ cm^{-3}) limit to this argument see Sh. M. Kogan, Sov. Phys. Semicond. 7, 828 (1973), and ref. 1.

18. R. N. Hall and T. J. Soltys, IEEE Trans. Nucl. Sci. NS-18, No. 1, 160 (1971).

19. M. S. Skolnick, L. Eaves, R. A. Stradling, J. G. Portal, and S. Askenazy, Solid State Commun. 15, 1403 (1974).

20. L. S. Darken and S. A. Hyder, submitted for publication.

21. E. M. Bykova, T. M. Lifshits, and V. I. Sidorov, Sov. Phys. Semicond. 7, 671 (1973).

22. E. E. Haller, W. L. Hansen, and F. S. Goulding, IEEE Trans. Nucl. Sci. NS-22, No. 1, 127 (1975).

23. E. M. Bykova, M. I. Iglitsyn, E. A. Kurkova, D. I. Levinzon, V. I. Sidorov, and V. A. Shersdel, Industrial Laboratory 42, 554 (1976).

23a. C. H. Burton, J. Electrical and Electroncis Engineering, Australia 1, 14, (1981).

24. H. W. H. M. Jongbloets, J. H. M. Stoelinga, M. J. H. van de Steeg, and P. Wyder, Physica 89B, 18 (1977).

25. R. N. Hall, Inst. Phys. Conf. Ser. 23, 190 (1975).

26. R. N. Hall and T. J. Soltys, IEEE Trans. Nucl. Sci., NS-25, No. 1, 385 (1978).

27. E. E. Haller, Phys. Rev. Lett. 40, 584 (1978).

28. E. E. Haller, B. Joos, and L. M. Falicov, Phys. Rev. B 21, 4729 (1980).

29. B. Joos, E. E. Haller, and L. M. Falicov, Phys. Rev. B 22, 832 (1980).

30. E. E. Haller and L. M. Falicov, Phys. Rev. Lett. 41, 1192 (1978).

31. For a review of the optical characterization of deep levels, see B. Monemar and H. G. Grimmeiss, Prog. Crystal Growth Charact. 5, 47 (1982).

32. E. E. Haller, G. S. Hubbard, W. L. Hansen, and A. Seeger, Inst. Phys. Conf. Ser. 31, 309 (1977).

33. R. Baron, M. H. Young, J. K. Neeland, and O. J. Marsh in: Semiconductor Silicon - 1977 (The Electrochemical Society, New Jersey 1977) p. 367.

34. E. E. Haller, P. P. Li, G. S. Hubbard, and W. L. Hansen, IEEE Trans. Nucl. Sci. NS-26, No. 1, 265 (1979).

35. For a more complete treatment of DLTS, see G. L. Miller, D. Lang, and L. C. Kimerling, Ann. Rev. Mat. Sci. 7, 377 (1977).

36. A. O. Evwaraye, R. N. Hall, and T. J. Soltys, IEEE Trans. Nucl. Sci. NS-26, No. 1, 271 (1979)

37. G. L. Miller, J. V. Ramirez, and D. A. H. Robinson, J. Appl. Phys. 46, 2638 (1975).

38. R. N. Hall, Inst. Phys. Conf. Ser. 23, 190 (1975).

39. For further evidence of Cu-related deep level contamination in high-purity germanium see also G. S. Hubbard, this volume.

40. R. H. Pehl, N. W. Madden, J. H. Elliott, T. W. Raudorf, R. C. Trammell, and L. S. Darken, Jr., IEEE Trans. Nucl. Sci. NS-26, No. 1, 321 (1979).

41. J. Guldberg, Appl. Phys. Lett. 31, 578 (1977).

42. C. Kim, H. Kim, A. Yusa, and S. Miki, IEEE Trans. Nucl. Sci. NS-26, No. 1, 292 (1979).

43. K. Graff and H. Pieper in: Semiconductor Silicon - 1981 (The Electrochemical Society, New Jersey 1981) p. 331.

44. E. Weber and H. G. Riotte, Appl. Phys. Lett. 33, 433 (1978).

45. H. J. Rijks, J. Bloem, and L. J. Giling, J. Appl. Phys. 50, 1370 (1979).

46. K. Graff and H. Pieper, J. Electrochem. Soc. 128, 669 (1981); P. J. Ward, J. Electrochem. Soc. 129, 2573 (1982).

47. See, for example, A. S. Grove, Physics and Technology of Semiconductor Devices (John Wiley and Sons, New York 1967).

48. S. A. Hyder, EG&G ORTEC, Oak Ridge, TN, private communication.

49. These effects have been discussed by P. A. Tove, S. A. Hyder, and G. Susila, Solid State Electron. 16, 513 (1973).

50. R. N. Hall, Appl. Phys. Lett. 29, 202 (1976).

51. For a discussion of the problems encountered in drifting lithium through germanium, see Semiconductor Nuclear Particle Detectors and Circuits, W. L. Brown, W. A. Higinbotham, G. L. Miller, and R. L. Chase, eds. (National Academy of Science, Washington 1969) p. 207.

52. See also P. Glasow, this volume.

53. For example, see L. Jastrzebski and P. Zanzucchi in: Semiconductor Silicon - 1981 (The Electrochemical Society, New Jersey 1981) p. 138.

54. For a review of these defects, see A. J. R. de Kock in Crystal Growth and Material, E. Kaldis and H. J. Scheel, eds. (North-Holland, New York 1977) p. 662.

55. R. L. Marchand, A. R. Stivers, and C. T. Sah, J. Appl. Phys. 48, 2576 (1977).

56. E. E. Haller, Inst. Phys. Conf. Ser. 46, 205 (1979).

57. S. J. Pearton, Appl. Phys. Lett. 40, 253 (1982).

58. R. C. Frank and J. E. Thomas, J. Phys. Chem. Solids 16, 144 (1960).

59. W. L. Hansen and E. E. Haller, IEEE Trans. Nucl. Sci. NS-21, No. 1, 251 (1974).

60. Oxygen related thermal donors, which would be generated by this thermal sequence, have been observed only in oxygen-doped germanium $[O] > 10^{16}$ cm^{-3}. See, for example, P. Clauws, J. Broeckx, E. Simoen, and J. Vennik, Solid State Commun., to be published.

61. R. J. Fox, IEEE Trans. Nucl. Sci. NS-13, No. 1, 367 (1966).

62. L. S. Darken, Jr., J. Electrochem. Soc. 126, 827 (1979), L. S. Darken, Jr., J. Electrochem. Soc, 129, 226 (1982).

63. H. Reiss, C. S. Fuller, and F. J. Morin, Bell Syst. Tech. J. 25, 535 (1956).

64. F. A. Kroger, The Chemistry of Imperfect Crystals (North-Holland, Amsterdam 1964).

65. H. Reiss and C. S. Fuller, Trans. Am. Inst. Min. Metall. Pet. Eng. 206, 276 (1956).

66. L. S. Darken, IEEE Trans. Nucl. Sci. NS-26, No. 1, 324 (1979).

67. K. Fitzner, K. T. Jacob, and C. B. Alcock, Metal. Trans. B 8B, 669 (1977).

68. S. Otsuka, T. Sano, and Z. Kozuka, Metal. Trans. B 12B, 427 (1981).

69. W. Kaiser and C. D. Thurmond, J. Appl. Phys. 32, 115 (1961).

70. E. J. Millett, L. S. Wood,and G. Bew, Br. J. Appl. Phys. 16, 1593 (1965).

70. W. D. Edwards, J. Electrochem. Soc. 115, 753 (1968).

71. Y. Yatsurugi, N. Akiyama, Endo and T. Nozaki, J. Electrochem. Soc. 120, 975 (1973).

72. R. N. Hall, IEEE Trans. Nucl. Sci. NS-19, No. 1, 266 (1972).

73. For a review of the present status of experimental methods and results on the electrical properties of dislocations see H. J. Queisser in: Defects in Semiconductors, S. Mahajan and J. W. Corbett eds. (North-Holland, New York 1983). See also Hunfeld Symposium: J. de Physique 40, Suppl. C6 (1979).

74. W. Shockley, Phys. Rev. 91, 228 (1953).

75. W. T. Read, Phil. Mag. 45, 775 (1954); 45, 119 (1954); 46, 111 (1955).

76. I. L. F. Ray and D. J. H. Cockayne, Proc. Roy. Soc. A32, 593 (1971).

77. G. Packeiser and P. Haasen, Phil. Mag. 35, 821 (1977).

78. S. Marklund, Phys. Stat. Sol.(b) 92, 83 (1979).

79. R. Jones, J. de Physique 40, Suppl. C6, 33 (1979).

80. W. Schroter, E. Schiebe and H. Schoen, J. of Microscopy 118 pt. 1, 22 (1980).

81. L. C. Kimerling and J. R. Patel, Appl. Phys. Lett. 34, 73 (1979).

82. J. R. Patel and L. C. Kimerling in: Defects in Semiconductors, J. Narayan and T. Y. Tan eds. (North-Holland, New York 1981) p. 273.

83. H. Schoen, Doctoral Thesis, Gottinger (1979). See also Ref. 80.

84. A. Ourmazd and G. R. Booker, Phys. Stat. Sol.(a) 55, 771 (1979).

85. R. Labusch and W. Schroter in: Dislocation in Solids, F. R. N. Nabarro ed. (North-Holland, New York).

86. P. A. Glasow and E. E. Haller, IEEE Trans. Nucl. Sci. NS-23, 92 (1976).

87. G. S. Hubbard, E. E. Haller, and W. L. Hansen, IEEE Trans. On Nucl. Sci. NS-26, No. 1, 303 (1979).

88. G. S. Hubbard and E. E. Haller, J. Electron. Mater. 9, 51 (1980).

89. W. Schroter, J. de Physique 40, Suppl. C6, 51 (1979).

90. J. R. Patel and L. C. Kimerling in: Defects in semiconductors. J. Narayan and T. Y. Tan eds. (North-Holland, New York 1981) p. 273.

91. D. B. Holt, J. de Physique 40, Suppl. C6, 189 (1979).

92. A. G. Tweet, Phys. Rev. 106, 221 (1957).

93. A. G. Cullis and L. E. Katz, Phil. Mag. 30, 1419 (1974).

94. C. Bull, P. Ashburn, G. R. Booker, and K. H. Nicholas, Solid State Electron. 22, 95 (1979).

95. T. Koji, W. F. Tseng and J. W. Mayer, Appl. Phys. Lett. 32, 749 (1978).

96. C. J. Varker and K. V. Ravi, J. Appl. Phys. 45, 272 (1974).

97. K. V. Ravi, C. J. Varker, and C. E. Volk, J. Electrochem. Soc. 120, 533 (1973).

98. R. N. Hall, General Electric, Schenectady, NY, private communication.

99. R. D. Westbrook, Nucl. Instr. and Meth. 108, 335 (1973).

100. H. H. Busta and H. A. Waggener, J. Electrochem. Soc. 124, 1424 (1977).

101. A. Goetzberger and W. Shockley, J. Appl. Phys. 31, 1821 (1960).

102. S. T. Hsu, R. J. Whittier, and C. A. Mead, Solid State Electron. 13, 1055 (1970).

103. T. Figielski, Solid State Electron. 21, 1403 (1978).

104. L. S. Darken, R. C. Trammell, T. W. Raudorf, R. H. Pehl, and J. H. Elliott, Nucl. Instr. and Meth. 171, 49 (1980).

105. A. Fong, J. T. Walton, E. E. Haller, H. A. Sommer, and J. Guldberg, submitted to Nucl. Instr. Meth.

106. J. T. Walton and E. E. Haller, this volume.

107. W. L. Hansen, this volume.

108. D. Itoh, I. Namba, and Y. Yatsurugi, this volume.

SCINTILLATOR MATERIAL GROWTH

O. H. NESTOR
The Harshaw Chemical Co., 6801 Cochran Road, Solon, Ohio, USA

ABSTRACT

Selected aspects of NaI(Tl), $Bi_4Ge_3O_{12}$ and $CdWO_4$ growth are addressed. Purity and growth control, dopant level effects, specific imperfections and plastic deformation in large scale commercial production are considered.

INTRODUCTION

The dominant inorganic scintillator materials today are NaI(Tl), $Bi_4Ge_3O_{12}$ and $CdWO_4$. The first of these, dating back to 1949 [1], is the industry standard for energy conversion efficiency. It has had wide and diverse application in nuclear research, medical diagnostics, geological exploration and other areas. BGO, i.e. $Bi_4Ge_3O_{12}$, developed in the mid-'70s [2], was quickly adopted for CT applications because of its moisture resistance, low afterglow and effective x-ray absorption (Table I). $CdWO_4$, first characterized as a scintillator in 1949 [4], combines the favorable properties of BGO with better light output and, under recent development, has become the material of choice in CT applications. In turn, BGO is now favored for high energy physics research by virtue of lower fluorescent decay constant and greater absorption coefficient.

There are a number of papers on the growth and properties of the above materials. It is not intended that these should be reviewed here. Rather, there are little-mentioned aspects of the large scale commercial production of these materials for scintillation applications that deserve comment. Each

Table I.
Scintillator characteristics under x- or γ-ray excitation [3].

	NaI(Tl)	BGO	$CdWO_4$
Crystal System	Cubic	Cubic	Monoclinic
Density (gm/cm^3)	3.67	7.13	7.90
150 KeV absorption coef. (cm^{-1})	2.2	10.0	7.7
Light output (relative)	100	10	40
Peak wavelength (nm)	415	480	540
Primary decay constant (nsec)	230	300	5000
Afterglow (% residual after 3 millisec)	0.3-5.	.005	.005

of these materials is technically interesting from a material preparation point of view either by way of posing difficulties that cry for solution or allowing an ingenious route to a desired end result. Such topics constitute the focus of this review.

A thread common to the materials discussed here is the approach to crystal growth. All are congruently melting materials that solidify into a crystallographic structure that is stable down to room temperature. Hence they are all grown from the melt, a near-requirement where high production rates are to be sustained.

THALLIUM-ACTIVATED SODIUM IODIDE

In pure form, sodium iodide is an inefficient scintillator. Activated with Tl^{+} in the range of 0.02 -0.2 mole percent, it has been one of the most important scintillation phosphors. This material is typically grown by the Stockbarger process without seeding, i.e. with growth nucleated spontaneously. Crystals ranging up to 30 inches in diameter x half ton mass, as illustrated in Figure 1, have been grown. Such crystals are not truly single, but rather contain five to ten components as delineated by small angle grain boundaries that crop out at surfaces or evident under strong illumination by the scattering effected by impurities on the boundaries. Such boundaries occur in fewer numbers as the crystal diameter decreases.

Grain boundaries generally do not impact on the utility of NaI(Tl) except for camera plate applications wherein they may mar the imaging processes. Material for camera plates is hence selected for minimal boundary decorations. This is done in lieu of a difficult and costly alternative of evolving a process that would yield boundary-free large crystals. Boundaries are tolerated in commercial production of large crystals as a concession to demands for high throughputs and reasonable market prices.

Starting materials for growth of NaI(Tl) are produced in quarter-ton batches to stringent specifications relating to anionic as well as cationic impurities. Specifications were derived by characterizing NaI(Tl) crystals doped with individual impurities in terms of the properties relevant to scintillation: color, resistance to color change under UV-excitation, afterglow, and light generation under nuclear radiation. In addition, reactive chemistry is effected in the melt phase to further enhance crystal quality. A recent patent issue [5] teaches the use of a borate-silicate addition to the growth stock to getter trace impurities in the melt. The reaction products are insoluble impurity silicate and impurity borate particles which tend to agglomerate and characteristically, by attachment to the crucible walls or by rejection to the last grown portion of the crystal, come to lodge at the periphery of the crystal, whence they can be discarded. Excess unreacted silica is insoluble in the NaI melt and collects along with the impurity agglomerate, or floc. The floc may not contain all of the added borate; in such case there is borate in solid solution in the crystal, but this has been found not to impair the material's vital properties.

While gettering reduces the level of undesired impurities in the melt, it takes up some of the activator as well. However, the relative amounts of getter and activator are sufficiently disparate that the required activation of the crystal is insured.

The activation desired in NaI(Tl) is the thallous (Tl^+) ion entering substitutionally into the NaI lattice in the amount of 0.02-0.2 mole percent Tl. The height of the photopeak generated under γ-excitation varies little for this thallium range, as illustrated in Figure 2. This is a fortuitously favorable result, since the segregation coefficient for Tl^+ in NaI is 0.2 and therefore maintaining uniform thallium distribution throughout a crystal is, in effect, not feasible. The thallium distribution illustrated in Figure 3 is rather typical of commercial production when some effort is made to limit non-uniformity. Now the value of the response curve shown in Figure 2 is evident, for it promises reasonably uniform pulse height response, and hence acceptable resolution, even with non-uniformly doped NaI(Tl), if the Tl levels exceed 0.02 mole percent.

The pulse heights for the two sets of data shown in Figure 2 have been normalized to 100 at 0.04 mole percent Tl, when in fact the relative heights should differ set-to-set due to the different γ-ray energies entailed in generating the data. The superposition shown in the figure was effected to illustrate the similarity of results. The gradual decrease with thallium content above 0.06 mole percent Tl was not emphasized by the original investigators. The decrease may reflect an effect of Tl dimers whose existence in sensible concentrations for the Tl range considered (0.1 mole percent) was first advanced by Van Sciver [8].

There has been demand for large NaI(Tl) structures that would be economically prohibitive, if it were required to produce them from single crystal material. This demand has been met by plastically deforming NaI(Tl) crystals after growth i.e. by hot-pressing or -extruding them to new shapes commensurate with the desired end geometry, while retaining transparency and performance characteristic of single crystals. The long cylinders of several different cross-sections shown in Figure 4 are NaI(Tl) articles produced by extrusion; from these forms they are finished into detectors for aerial geological surveys, portal monitors, etc. Figure 5 shows a hemisphere of the SLAC crystal ball comprised of some 350 pyramidal crystals of NaI(Tl). These were cut from "fortrusions", the products of hot-pressing crystals into a mold of such geometry that the desired pyramids (seven dimensionally different units were needed) could be cut therefrom with minimal waste of material.

The theoretical basis on which the forging/extrusion technology rests is given by the von Mises criterion [9], specifying that five independent shear systems are required in the general case to support plastic deformation, to permit a grain to conform to the distortion of its neighbors. This translates into five independent glide or slip systems for crystals such as sodium iodide. NaI has two primary glide systems of the form $\{110\}\langle 110\rangle$ and three secondary glide system of the form $\{001\}\langle 110\rangle$ active simultaneously at temperatures below melting.

The preceding illustrations of deformation in NaI(Tl) refer to shape control. This impacts on costs through efficient utilization of material as well as permitting economies in fabrication. The deformation can also be effected so as to produce a stronger structure. Strength enhancement is determined by grain size and this is controllable through parameters of deformation rate and temperature. Figure 6 illustrates a fine grain structure achieved in a hot-pressing experiment; the structure is essentially equiaxed and of approximately 20 micron average grain size. The yield strength of the article represented here was measured to be 1700 psi in tension. Single crystal NaI(Tl) has tensile yield strength of about 300 psi, while bars such as those shown in Figure 4 typically have had centimeter - size grains and tensile strength of 600-800 psi. Thus, a "new" structure is available in NaI(Tl) that may be put to use in generating scintillator detectors of enhanced properties, e.g. thinner camera plates to achieve better resolution in imaging applications.

There is an interesting comparison to be made between the above route to a transparent polycrystalline scintillator and a pressed powder route. The latter circumvents crystal growth but then must deal with scatter at grain boundaries due to impurities and gases included thereon. The former contends with the effort to grow a crystal, but since that process results in minimal gas and impurity content, the conversion to polycrystallinity is straightforward as to transparcency.

BISMUTH ORTHOGERMANATE (BGO)

BGO is an intrinsic scintillator wherein the principle stopping power and the seat of fluorescence lie in one and the same atom, Bi, having high atomic number (Z=83). Therefore, acivator distribution and sensitizing effects are not applicable to this material.

BGO is typically grown utilizing the Czochralski technique. The melt is very viscous and exhibits supercooling to an extreme; melts have been observed to persist in excess of 100`C below the freezing point. Large crystals today measure 3" diameter x 9" length. The principal deterrent to getting larger crystals is the incidence of a defect described below.

The major problem encountered in the growth of BGO in ambient air is one that has been labeled "coring." The nomenclature refers to a defect that normally crowds the axial zone of the as-grown crystal, or boule. The defect is a distribution of particulate inclusions, as shown in Figure 7. A number of these exhibit elongation in the growth direction implying migration toward the hottest part of the growth chamber. In general, the morphology of the inclusions is not that of voids or gaseous inclusions and in fact the inclusions usually exhibit straw coloration. One of the inclusions is shown under high magnification in Figure 8a along with SEM micrographs showing the distribution of x-rays characteristic of germanium (Figure 8b) and bismuth (Figure 8c) generated by the rastered electron beam of the instrument. It is evident that the inclusion is an off-stochiometry globule of material, deficient in germanium and containing excess bismuth relative to the lattice. The Bi/Ge ratio of other inclusions that have been examined is similarly distorted. The role of the platinum particle evident in Figure 8d relative to the globular inclusion is unknown.

When crystals free of such inclusions were utilized as growth stock, the resultant boule was found to exhibit coring to a high degree. This suggests that the incidence of globules is germane to the BGO melt, perhaps related to its highly viscous nature, whereby thorough mixing is elusive, or to pre-freezing clustering phenomena in the viscous melt [10]. The possible role of impurities in nucleating such clusters is an unknown. While better understanding of this problem is sought, recourse to growth technique has been effective in providing some control on the incidence of inclusions. With such technique it has been observed, in concurrence with results reported by others [11], that overall quality is improved and the incidence of coring is lessened by regrowth of crystals [12]. This indicates that impurities still limit quality and may play a role in nucleating the off-stoichiometric inclusions discussed here.

Deforming BGO plastically by hot-pressing or -extrusion has not been successful even on approaching the melting point to within 40°C.

CADMIUM TUNGSTATE

$CdWO_4$ is an intrinsic scintillator whose fluorescence under x- and γ-ray excitation has been assigned to a tungsten-oxygen complex [13,3]. $CdWO_4$ is grown at Harshaw by the Czochralski technique from growth stock synthesized from basic components and purified to insure proper purity level and control. Significant vaporization occurs during growth, representing loss of CdO principally, but also some tungstic oxide. Specific precautions are taken to protect personnel against exposure to Cd vapors designated as being carcinogenic.

The growth of $CdWO_4$ could be quite difficult except for the adaptation of automatic control systems to the growth process. Requirements for controlled growth are understandable in terms of the standard phase diagram for the WO_3 - CdO system (Figure 9). With the preferential loss of CdO by vaporization during growth and with solidification of the stoichiometric composition further accelerating depletion of CdO from the melt, the liquidus curve must be followed in the direction indicated by the arrow, i.e. toward the WO_3 rich side of the compound stoichometry. Twenty years ago this would be effected by constant manual adjustment of power to the growth station, based on continuous observation of the growth meniscus, so as to keep abreast of the change in instantaneous melting point with time. This generally would entail several trials to zero-in on a suitable program, a procedure to be repeated as changes in furnace configuration are made. Today with automatic controls, whether based on instrumental observation of the meniscus or sensing of melt temperature near the crystal periphery or on continuous weighing of either the crystal or the crucible plus melt during growth, tracking the instantaneous melting point requirement has become a trivial matter.

A typical $CdWO_4$ growth result is illustrated in Figure 10, showing a longitudinally cleaved section of an (001) axis boule. There are no inclusions evident, no coring observed, until late in the growth. Analytically, the zones of inclusions were identified to be $CdWO_4$ plus WO_3, the latter component accruing as a result of excess WO_3 and eutectic buld-up in the melt. Coring along the growth axis in the form described by Robertson, Young and Telfer [15] is not typically observed. It is suggested that this difference is attributable to details of the growth technique.

SUMMARY

Selected aspects of the production of NaI(Tl), $Bi_4Ge_3O_{12}$ and $CdWO_4$ have been addressed. The first of these relies on dopant activation whereas the others are intrinsic scintillators. All are grown under controlled solidification from their melts.

Very large crystals of NaI(Tl) are produced commercially utilizing reactive melt chemistry as a final control on impurities. Getting uniform dopant activation is not feasible, but fortuitously there is a broad range of Tl^+ activation within which pulse height under -ray excitation varies but little. Hot-pressing and -extrusion is a convenient adjunct to crystal growth in processing NaI(Tl), utilized to attain crystal size and shape that are otherwise not accessible on a practical basis, to promote economic utilization of material and/or to enhance material strength.

BGO growth is principally an exercise in controlling particulate inclusions of non-stoichiometric composition, constituting the present limit on the available crystal size. The origin of the inclusions is still speculative. It is suggested to lie in pre-freezing cluster phenomena in the melt, perhaps nucleated by impurities.

In light of normal vaporization losses, $CdWO_4$ must be regarded as entailing growth from a non-stoichiometric melt of constantly changing composition. Tracking the melting point to achieve uniform crystal diameter, commensurate with good quality, is reduced to a trivial problem with the diameter control technology now available. Crystal size is ultimately limited by the approach of the melt at the growth interface to a eutectic composition.

REFERENCES

1. R. Hofstadter, Phy. Rev. 74, 100 (1948); Phys. Rev. 75, 796 (1949).
2. O. H. Nestor and C. Y. Huang, IEEE Trans. Nucl. Sci. NS-22, 68 (1975).
3. M. R. Farukhi, Workshop, IEEE Trans. Nucl. Sci. NS-29, 1237 (1982).
4. R. H. Gillette, Rev. Sci Inst. 21, 294 (1950).
5. C. F. Swinehart, U. S. Patent No. 4,341,654 (Harshaw).
6. J. A. Harshaw, H. C. Kremers, E. C. Stewart, E. K. Warburton and J. O. Hay, USAEC Bulletin 40,N1577 (1952).
7. L. M. Belyaev, M. D. Galanin, Z. L. Morgenshtern and Z. A. Chizhikova, Proc. Acad. Sci. USSR 105 (1), 57 (1955).
8. W. Van Sciver, IEEE Trans. Nucl. Sci. NS-13, 138 (1966).
9. See M. T. Sprackling, "The Plastic Deformation of Simple Ionic Crystals" (Academic Press, 1976) p. 168.
10. A. R. Ubbelohde "The Molten State of Matter" (John Wiley and Sons, 1978), 342 et seq.
11. K. Takagi, T. Oi, T. Fukazawa, M. Ishii and S. Akiyama, (Hitachi Chem. Co.), publication pre-print.
12. J. R. Hietanen, Harshaw Chemical Co., private communication.
13. M. J. J. Lammers, G. Blasse and D. S. Robertson, Phys. Stat. Sol. 63(a) 569 (1981).
14. I. P. Kislyakov and B. P. Lopatin, Russ. J. Inorg. Chem. 12(11), 1674 (1967). [See E. M. Levin and H. F. McMurdie, Phase Diagrams for Ceramists (Amer. Ceram. Soc., Columbus, Ohio, Publ.), Figure 4323].
15. D. S. Robertson, I. M. Young and J. R. Telfer, Jour. Mat'l Science 14, 2967 (1979).

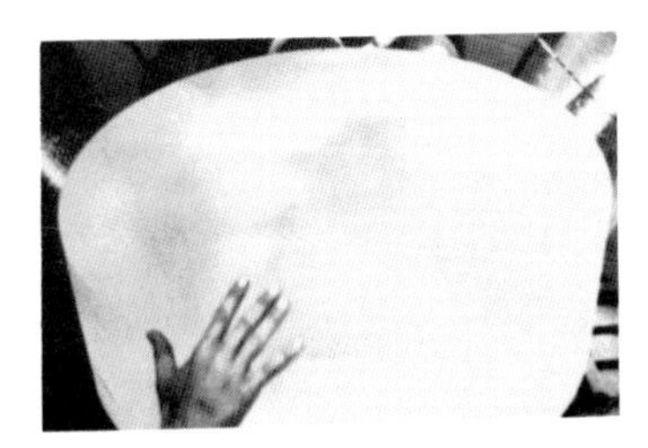

Fig. 1. 30 inch diameter Stockbarger-grown crystal.

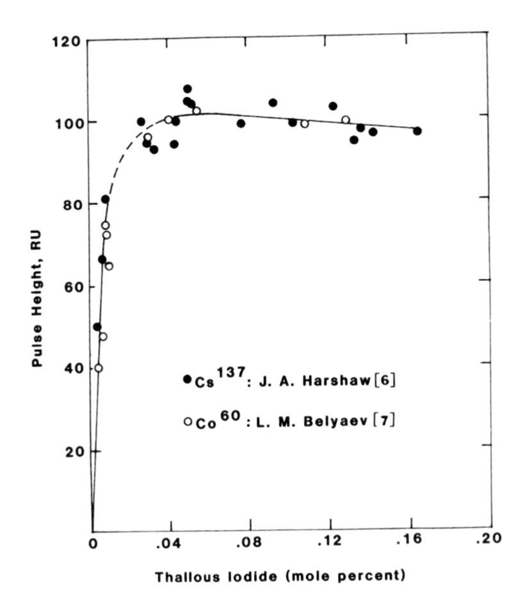

Fig. 2. Dependence of pulse height on thallium activation in sodium iodide under γ-ray excitation.

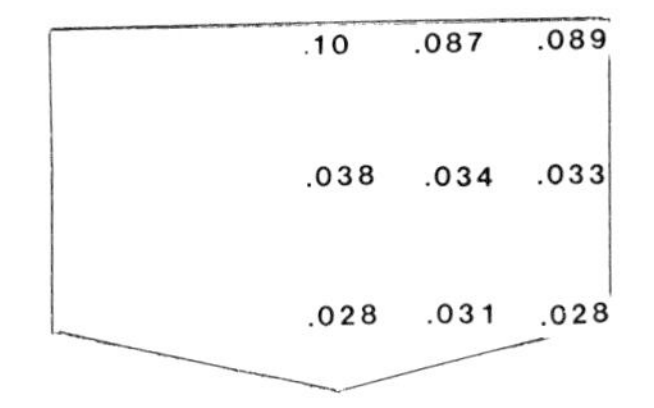

Fig. 3. Thallium distribution in a 19 inch diameter NaI(Tl) crystal (mole percent).

Fig. 4. NaI(Tl) extrusions.

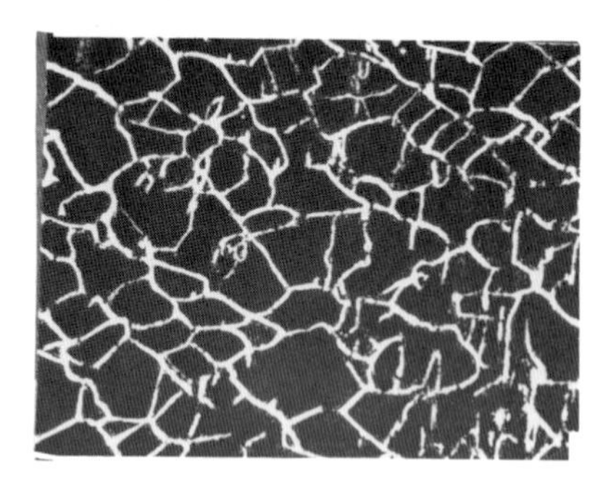

Fig. 6. Fine-grain (20 micron average) structure in hot-pressed NaI(Tl).

Fig. 5. Section of SLAC ball.

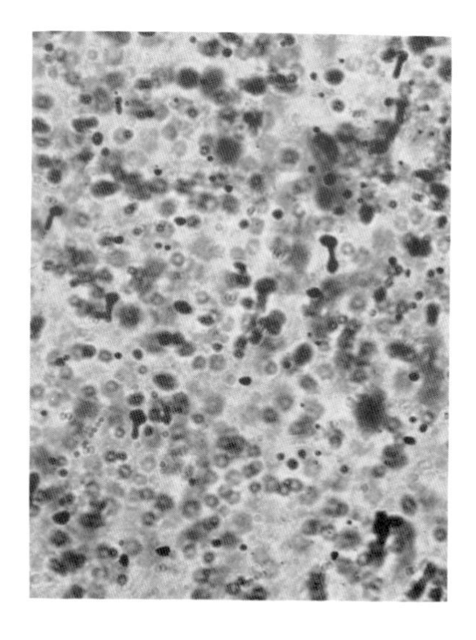

Fig. 7. Particulate inclusions in BGO. The longest in this view measures 0.1mm.

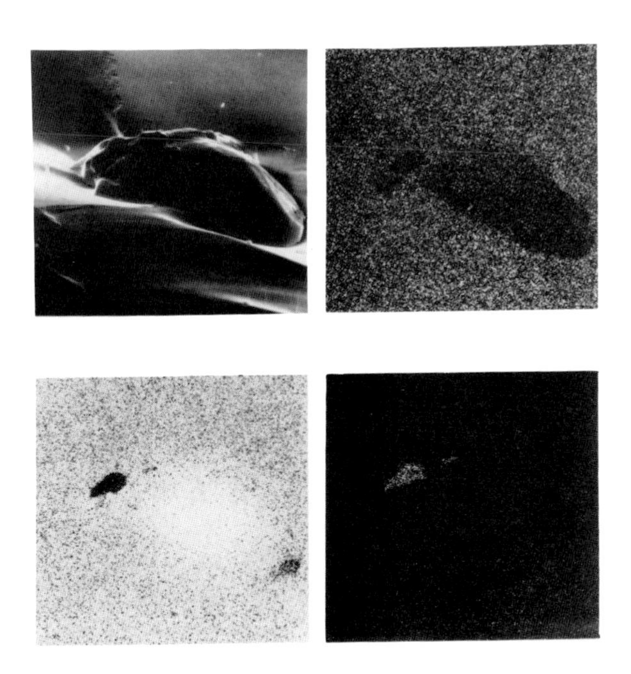

Fig. 8. Inclusion in BGO. (a) optical image; (b)-(d) x-ray "images" recorded via scanning electron microscopy.

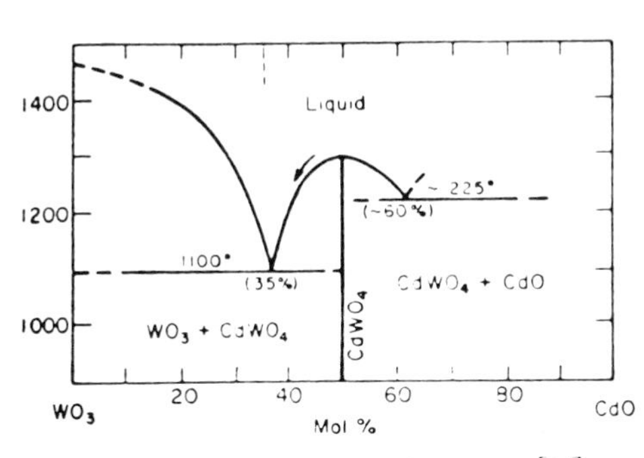

Fig. 9. WO_3-CdO phase diagram. [14]

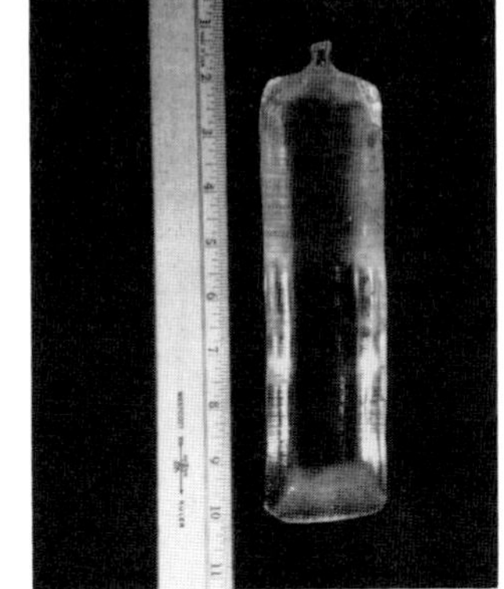

Fig. 10. Czochralski-grown $CdWO_4$ crystal.

CADMIUM TELLURIDE DETECTORS AND APPLICATIONS

PAUL SIFFERT
Centre de Recherches Nucléaires, groupe de Physique et Applications des Semiconducteurs (PHASE), 67037 Strasbourg-Cedex FRANCE

ABSTRACT

The current possibilities as well as the limitations today of cadmium telluride photon counters and spectrometers are reviewed. The various types of detectors are considered together with their main characteristics. These detectors are compared with mercuric iodide counters and possible applications of these devices are explained.

INTRODUCTION

About fifteen years ago, the germanium and silicon solid state detectors strongly improved the possibilities of X and γ -ray spectroscopy. In experiments, in which high energy resolution was requested, these detectors definitively surpassed the conventional photomultiplier-scintillator systems. The need of cooling some detectors, (down to LN_2 temperature in the case of Ge (Li) or H-P-Ge), may constitute a real handicap in many applications; thus large scale use around reactors, in industry or in nuclear medicine can be limited by this requirement. Furthermore, both silicon and germanium are low atomic number Z (14, 32) materials, which is not best suited for nuclear photon spectroscopy.

These various reasons motivated a few groups around the world to try to develop a high efficiency spectrometer able to operate at room temperature with good energy resolution. Among the different semiconductors having the following characteristics : high Z, good carrier transport properties, sufficiently large (1.5 to 2 eV) bandgap to reduce thermal noise, only a few compounds can be considered (1), a listing of the potential detector material is shown on Table I. Today, only two of these compounds CdTe and HgI_2 have really been investigated as spectrometers. Furthermore, GaAs, which has the same Z as Ge has been considered, since its bandgap would allow a room temperature operation.

The goal of this paper is to review the current status of GaAs and CdTe detectors, HgI_2 being considered by VAN DEN BERG and SCHNEPPLE at this conference. Several reviews are also available in the literature (2-10).

GaAs DETECTORS

This III-V semiconductor was first considered, around 1960, for nuclear detector manufacturing, starting with high resistivity compensated crystals (11-13). However, due to strong trapping effects the performance has been quite poor. New interest appeared when it

Mat. Res. Soc. Symp. Proc. Vol. 16 (1983) © Elsevier Science Publishing Co., Inc.

Material	Atomic Nr. Z	Bandgap E_g (eV)	Mobility (cm^2/V. s.) electrons	holes	Density g. cm^{-3}
Si	14	1.12	1900	500	2.33
Ge	32	0.68	3800	1800	5.32
GaAs	31-33	1.45	8600	400	5.35
AlSb	13-51	1.62	200	700	4.25
GaSe	31-34	2.02	60	250	4.55
CdSe	48-34	1.74	50	50	
CdS	48-16	2.41	300	15	4.82
InP	49-15	1.35	4800	150	
ZnTe	30-52	2.25	350	110	
WSe_2	74-34	1.36	100	80	
BiI_3	83-53	1.70	680	20	
Bi_2S_3	83-16	1.3	1100	200	6.7
Cs_3Sb	55-51	1.6	500	10	
PbI_2	82-53	2.55	8	2	6.16
HgI_2	89-53	2.13	100	4	6.30
CdTe	48-52	1.45	1100	100	6.06

Table I
Potential semiconductor material, limited to binary compounds.

Table II : Main properties of the various CdTe type of crystals grown mainly today.

TECHNIQUE	COMPENSATION	RESISTIVITY (ohm. cm)	TYPE	MOBILITY (cm^2/V. s.)		LIFETIME. MOBILITY (cm^2 / V.)	
				electrons	holes	electrons	holes
Zone melting	no	100-500	N	1100			
Travelling heater method (THM)	no	10^4-10^7	P	700-900	70-85	3. 10^{-3}	3. 10^{-4}
	chlorine	10^7-10^9	P	700-900	70-85		
Bridgmann	indium	10^7-10^9	P	700-900	70-85		
Solvent	chlorine indium	10^8-10^9	P	700-900	70-85	6. 10^{-4}	5. 10^{-6}

	Li	Na	Cu	Ag	Au	N	P	As
CdTe	57.8	58.8	147	108			60	
CdSe	109						120	750
CdS	165	169	1100	260			600	
ZnTe	60.5	62.8	148	121	277		63.5	79
ZnSe	114	126	650	430	550	100	85	110
ZnS	150	190	1250	720			550	

Table III
Ionization energies of various ACCEPTORS in Zn and Cd compounds (meV).

Ionization energies of various DONORS in Zn and Cd compounds (meV).

	B	Al	Ga	In	F	Cl	Br	I	Li_I
CdTe				14.5		14.1			13.9
CdSe									
CdS			33.1	33.8	35.1	32.7	32.5	32.1	28
ZnTe		18.5				20.1			
ZnSe	25.6	25.6	27.2	28.2	28.2	26.2			21
ZnS		100							

became possible to grow high purity ($n = 6.0 \times 10^{12}$ cm^{-3}) epitaxial films either by liquid or vapor phase epitaxy (14-18) on impure GaAs or even Si : the stoichiometric defects being compensated with Fe. Resolutions as good as 2.95 keV have been achieved for 122 keV γ-rays on Schottky diodes about 60-80 μm thick and 1.5 mm in diameter. Unfortunately high quality epi-layers cannot exceed 100 μm in thickness. therefore, the efficiency of these devices remains small and no real progress has been achieved recently.

CdTE DETECTORS

Starting Materials. It should be emphasized that the development of high quality crystals of compound semiconductors is generally more delicate than that of Si or Ge. The reasons for this are the binary nature of the material, in which any deviation from perfect stoichiometry affects strongly the electrical as well as the transport properties, and also the easy decomposition of the compound when heated (for example for crystal growth). This high sensitivity among growth conditions leads to crystals having properties that can be quite different. These experimental difficulties explain partly the apparent low progress noted on these counters; furthermore these materials do not represent the same economical importance as Si or Ge.

Since the photon counters need rather thick sensitive zones, only the growth methods leading to high resistivity crystals are of real interest. Even if many techniques have been considered in the past (19), essentially three groups of methods are employed today: zone melting, travelling heater method (THM) and Bridgmann from normal freezing of a tellurium rich solution. Table II summarizes the major characteristics of the crystals. It appears that with the exception of the zone melting technique, which gives crystals of high quality, but low resistivity, just sufficient for X-ray spectroscopy, all the other methods operate in tellurium solvent, in order to reduce the growth temperature to about 700 to 900°C rather than to go through CdTe melting (1050°C). This temperature reduction has a decisive influence on the material pollution by impuri-

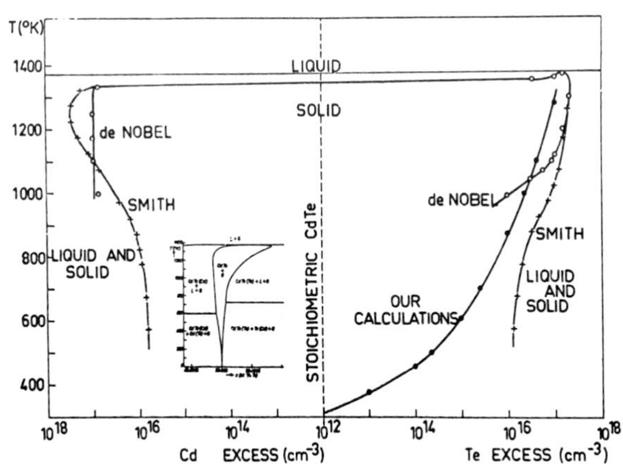

Fig. 1. Temperature T, composition x projection of phase diagram near stoichiométry. Inset gives the full diagram.

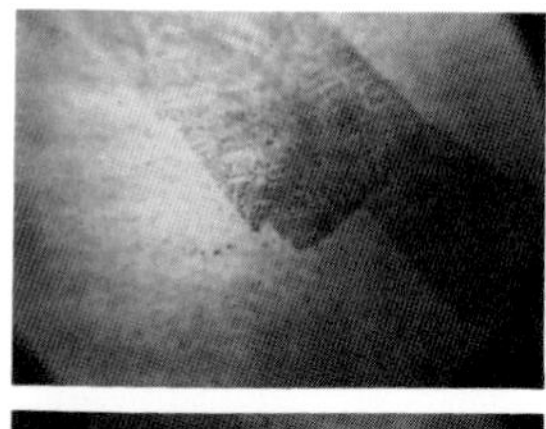

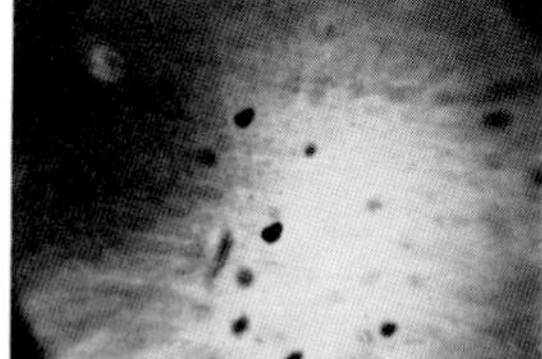

ties from quartz and furnace. The existence domain of the compound (solidus line) on the tellurium rich side of the temperature -composition projection of the phase diagram is reported in Fig. 1. This curve is strongly retrograde and when going from the growth temperature down to the ambient, the concentration of Te has to be reduced by several orders of magnitude. This can produce tellurium precipitations if the conditions are not controlled strictly (Fig. 2).

Fig. 2. Two kinds of Te precipitation in CdTe grown in Te.

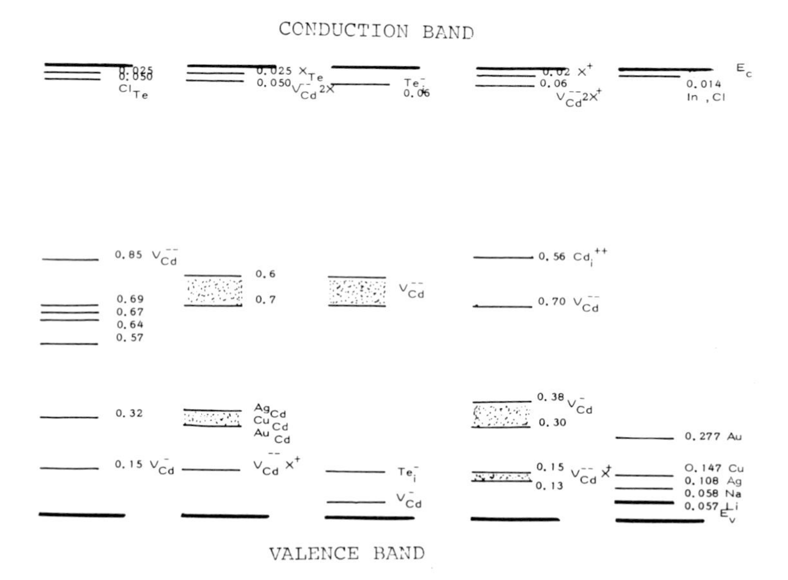

Fig. 3. Various models for impurity levels proposed for high resistivity CdTe.

Concerning the resistivity of these crystals, it has been shown that it is, theoretically, possible to reach very high resistivity (10^8 Ω.cm) without any external compensation. In fact, single charged cadmium vacancy V^{-}_{Cd}, doubly charged cadmium vacancy V^{--}_{Cd}, single charged cadmium interstitial Cd^{+}_{i} and doubly charged cadmium interstitial are present, which have been considered to give rise to several levels in the forbidden gap (Fig. 3) and have to be compensated either by indium, or mainly by halogens, essentially chlorine. The generally accepted model (20-22) gave to chlorine a similar role as lithium has in Si or Ge compensation in the presence of residual acceptors, the compensation equations being :

$$Li \rightleftarrows Li^{+} + e^{-}$$
$$+ Zn^{--} \rightleftarrows (Li^{+}\ Zn^{-})^{-}$$
$$+ Li^{+} \rightleftarrows (Li^{+})_{2}\ Zn^{--})$$

neutral triplet

In a similar way, the doubly charged vacancy, V^{--}_{Cd} which acts as an acceptor in CdTe can be compensated for by a donor X from group III (Al) or group VII (Cl, Br, I), the equation being :

$$X \rightleftarrows X^{+} + e^{-}$$
$$+ V^{--}_{Cd} \rightleftarrows (X^{+}\ V^{-}_{Cd})^{-}$$
$$+ X^{+} \rightleftarrows ((X^{+})_{2}\ V^{--}_{Cd})$$

neutral triplet

This description of the role of chlorine as compensating the native defects neglects the presence of residual chemical impurities. Recently, it has been shown that both in III-V and II-VI compounds, many levels attributed initially to physical defects are in fact due to chemical impurities. A recent review by BHARGAVA (23) compares the role of some impurities in Zn and Cd chalcogenides; Table III summarizes the main results for some donors and acceptors. The most astonishing result about CdTe is the origin of the level at E_V + 0.15 eV, generally attributed to the $V_{Cd}X$ complex which is shown to be due to copper (24) ; other impurities like aluminium also play an important role.

What is then the role of chlorine ?

HOSCHL et al (25) have shown that the $\mu\tau$ product of chlorine compensated CdTe increases with chlorine concentration present in the material ; ARKADEVA et al (26) observed that this behaviour is true up to concentrations of chlorine of 4.1×10^{18} cm^{-3} and that at higher values the resistivity decreased and finally becomes n-type. Therefore, the first role of chlorine is to produce a purification of the material by the generation of complexes with residual chemical impurities ; especially alkalii may be involved (27). The doping effect of chlorine and therefore, its compensation effect only starts when the chemical purification role is fullfilled. Above the indicated concentration, the material shifts to lower resistivity and converts to n-type. On the high resistivity P-type crystals ARKADEVA et al (28) observed a quasi continuous exponential distribution of levels over the upper half of the bandgap.

It should be mentionned that these results were not found in such a definitive manner in the western laboratories, when starting with highly purified materials. Therefore, besides copper, always associated with tellurium, it is possible that the doping and compensating effect of chlorine starts at lower concentration in these materials. In other words, the chemical or electronic role of chlorine depends on the purity of the compound material in which this halogen is introduced.

Fig. 4. Evolution of the crystal size grown by THM at various levels 1 mm, 5 mm, 10 mm, 2 cm, 3 cm, 5 cm.

In conclusion to this section, it should be noticed that the fundamental understanding of CdTe is still under discussion. The crystallographic quality strongly improved in recent years, since large crystals are now obtained both by zone melting and solvent growth methods (Fig. 4) ; furthermore, the small angle deviations ofvarious domains within a single crystal have been strongly reduced.

Detectors. The crystals are sliced in 1-4 mm thick wafers and cleaned by conventional semiconductor procedures. Then, the handling depends upon the nature of the material and the kind of detector wanted. Essentially, three categories of devices can be prepared :

a) For low energy, high resolution X-ray spectrometry, low resistivity N-type samples, from zone melting grown crystals, are used and a gold Schottky barrier is realized on a bromine-methanol etched surface, having further received some "kitchen" type surface treatment (28-30). The active dots are below 6 mm in diameter in order to keep the diode's capacitance small.

b) For γ -ray spectroscopy, both THM and solvent grown crystals are used, with or without Cl or In compensation. Due to the rather long time needed for crystal growth, these methods offer the advantage of lower temperature, leading to much less contamination by the quartz ampoule.

Essentially 3 kinds of structures have been investigated :

- Schottky diodes, mainly on lapped surfaces, because the barrier height on etched wafers is generally below 1 eV, giving rise to excess current and noise (31-38). However, on these lapped surfaces, the active area diameter is limited to about 4-5 mm.
- Ohmic contacts on semi-insulating materials, produced by electroless gold (platinum) chloride deposition (39) on etched surface composition (40-42), therefore, the electroless contact deposition techniques are of interest, since the reaction proceeds significantly below the surface, thus avoiding many of the cited surface complications (43). However, besides the redeposition of tellurium, the formation of tellurides, a migration of Au (or Pt) occurs, which is investigated in a separate contributed paper to this conference. When increasing the bias voltage, the noise quickly increases, limiting the electrical field and, consequently, the charge collection efficiency. Strong current reductions are possible by cooling (Fig. 5).

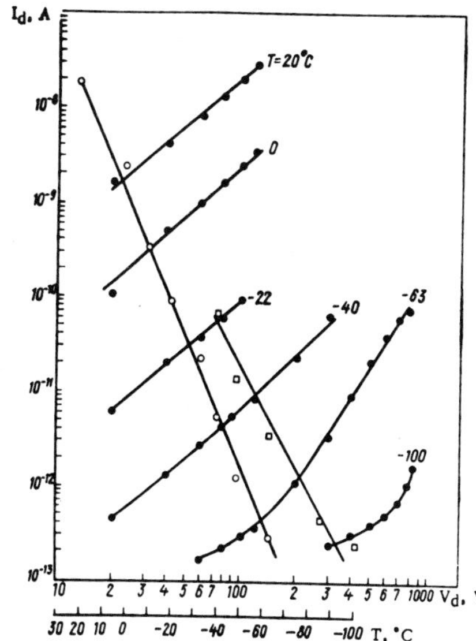

Fig. 5. Effect of temperature in I-V characteristics of Au-CdTe-Au structure. Contacts by $AuCl_3$.

- N-i-P structures would be best. They have also been considered but problems arise during the heat treatment for diffusion or annealing of damage after ion implantation (44,45). Perhaps pulsed annealing (46) can offer new possibilities even if the first results are not fully satisfying due to defects resulting from the high thermal gradient; much better results have been observed on ohmic contacts (47).

Up to recent time, the spectrometers prepared on crystals having a resistivity in excess of $\simeq 10^7$ Ω-cm as determined by the Van der Pauw technique exhibited strong polarization if rectifying contacts were used (see below) (48-50). This problem can now be solved (51).

c) For X-or γ-ray counting without energy resolution, CdTe can be used as solid ionization chamber. Generally, electroless contacts on etched surfaces are used as already indicated. More recently, for high flux measurements these counters have been employed in the photovoltaïc mode, without any external bias, like solar cells (52-53).

Electrical characteristics

a) I-V characteristic : both leakage current and breakdown voltage are strongly dependent on the nature of the material and the diode manufacturing process. In general, typical results are

as follow :
- on low resistivity N-type material with a gold Schottky barrier currents as low as 0.2 nA at 30 V and 3 nA at 100 V have been reached at room temperature. The breakdown voltage is close to 100-125 Volts.
- on high resistivity P-type crystals with lapped surfaces currents of 10^{-8} 10^{-9} A are obtained at 500-1000 V (3-4 mm diameter). The breakdown voltage exceeds 1000 V.
- on the same material but with electroless gold or platinum contacts, the current are higher by a factor 100 to 1000 and the breakdown is often around 100 V. This explains why these devices cannot be employed as spectrometers. In a recent paper, AGRINSKAYA and MATVEEV (54) found a correlation between the deviation from Ohm's law and resolution of detectors, that they attributed to homogeneity.

b) Capacitance : for diodes prepared on low resistivity material, authors have found (55-57) that the capacitance C changes with bias voltage V as : $C \alpha V^{-n}$ with $n \simeq 0.2$-0.3. This apparent deviation from a classical Schottky diode is due to the presence of a deep level at E_c - 0.55 eV (58).

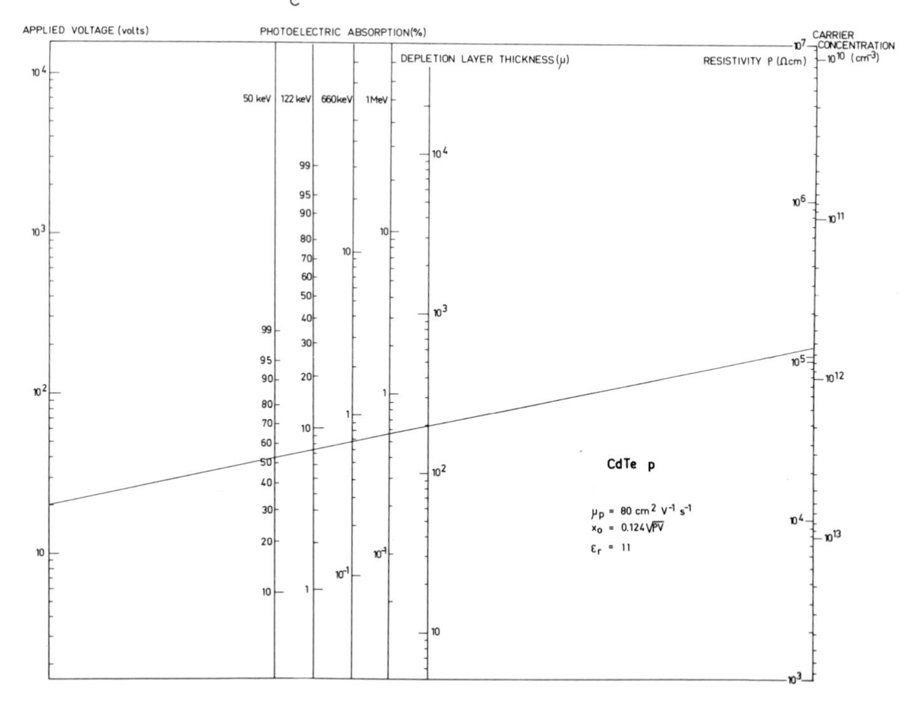

Fig. 6. Nomograph for the calculation of the depletion width and efficiency vs. resistivity, for P-type CdTe.

- For the high resistivity crystals, the capacitance is, for high bridge frequencies, independent of the voltage. For the undoped crystals, variations of C are observed with both temperature and frequency, due to the presence of a deep level at $E_V + 0.15$ eV. In chlorine compensated samples, a time dependent variation of C can be found under some particular conditions, due to polarization by a very deep level at about $E_V + 0.7$ eV.

c) Depletion layer width. Theoretically, the width of the depleted zone X under reverse bias V is expressed by :

$$X = 0.12\sqrt{\rho V} \text{ for P - type crystals (1) (Fig. 6)}$$
$$X = 0.4\sqrt{\rho V} \text{ for N - type crystals (2)}$$

In practice, the situation is, today, as follow :
- For diodes prepared on low resistivity (100-500 Ω.cm) N-type materials, the depletion width increases as expected up to a maximum value of about 100 μ , which gives an active volume of 10^{-4} cm^3, a value which constitutes a serious handicap for spectroscopy applications.
- For spectrometers realized on lapped high resistivity P-type materials, the depletion width is limited to a few mm the apparent resistivity ρ' of a biased diode has been found to be 500-1000 times lower than the value measured by the Van der Pauw method (59). This limitation of the extension of the depletion width probably has several origins, but two are dominant :

+ fast polarization appearing as soon as the bias voltage is applied (49) increasing the ionized $[N_A - N_D + N_T]$ concentration.

+ a high series resistance due to the lapped film (about 10μ m in thickness) reduces the effective applied voltage.

Therefore, the maximum effective depletion width is limited to 1 - 2 mm. This limit seems sufficient for most of the applications, since it is equivalent to more than 1 cm of germanium (photoelectric). Several procedures have been proposed to determine the effective extension of the space charge region (59-61). Since the active area is limited to about 5 mm in diameter, as already indicated, the volume of these spectrometers is also limited. Diode structures on etched materials are necessary for any further progress.
- For counters prepared on etched high resistivity P-type crystals on which the contacts are applied by electroless deposition of gold or platinum, the full space between electrodes is sensitive, by electrooptical measurements. Depletion zones up to 6-7 mm have been observed. However, since the breakdown voltage is rather small, the field within the material is weak, efficient charge collection can be achieved only near the negative biased electrode (Fig. 7) due to the difficulties holes have to reach the collecting electrode. As a result, strong degradation of the spectrometric properties appear when the thickness increases and in practice, the spectroscopic properties are poorer than on lapped structures.

Main detector properties

Pulse amplitude. The theoretical pulse amplitude is determined by the energy required for electron-hole generation (4.46 eV at

300 K). However, in practice, this amplitude is seen only for the purest low resistivity crystals. Essentially, two factors contribute to the reduction of the pulse amplitude :

- a small mobility-lifetime product $\mu\tau$, due to the presence of deep trapping centers. The best values, generally reported, are around 10^{-3} and 10^{-4} cm^2/V for electrons and holes, respectively, when deduced from the charge collection efficiency, from Hecht relationship. By using another approach based on the dependence of γ-ray counting rate vs applied voltage, KASHERININOV et al (61) indicated values as large as $5.\ 10^{-2}$ for electrons.
- surface preparation and contacts. Table IV gives the influence of surface conditions on pulse amplitude delivered for γ-rays and α-particles.

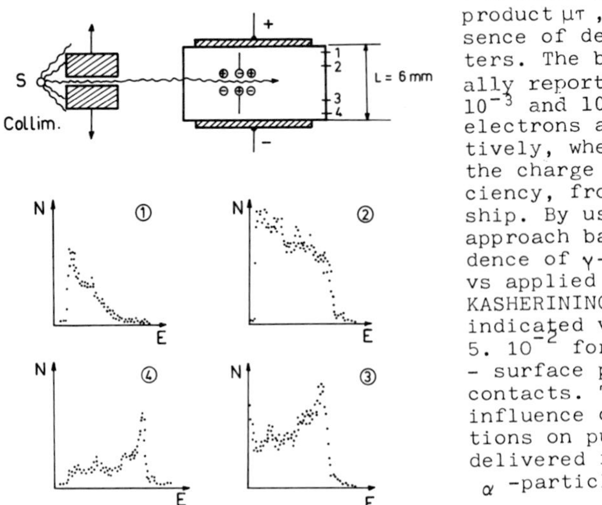

Fig. 7. Effect of point of irradiation on the shape of the γ-ray spectrum (^{57}Co). (27)

TABLE IV
Relative pulse amplitude for various surface treatment of CdTe

Nature of the radioactive source	Side of entrance of radiations	Lapped surface Aluminium contacts	Etched Surface Electroless Pt	Etched Surface Oxide + Al
alpha from ^{241}Am	+	0.35	0.20	0.85
	-	0	0.98	1
gamma from ^{57}Co	+	0.90	0.97	0.95
	-	0.90	0.99	0.97

The total charge collection efficiency η depends on the point x_o (measured from negative entrance electrode of the radiations) at which the electron-hole pair has been generated. For an uniform field its expression is given by :

$$\frac{\gamma_n}{d}\left[1 - \exp - \frac{(d-x_o)}{\gamma_n}\right] + \frac{\gamma_p}{d}\left[1 - \exp - \frac{x_o}{\gamma_p}\right]$$

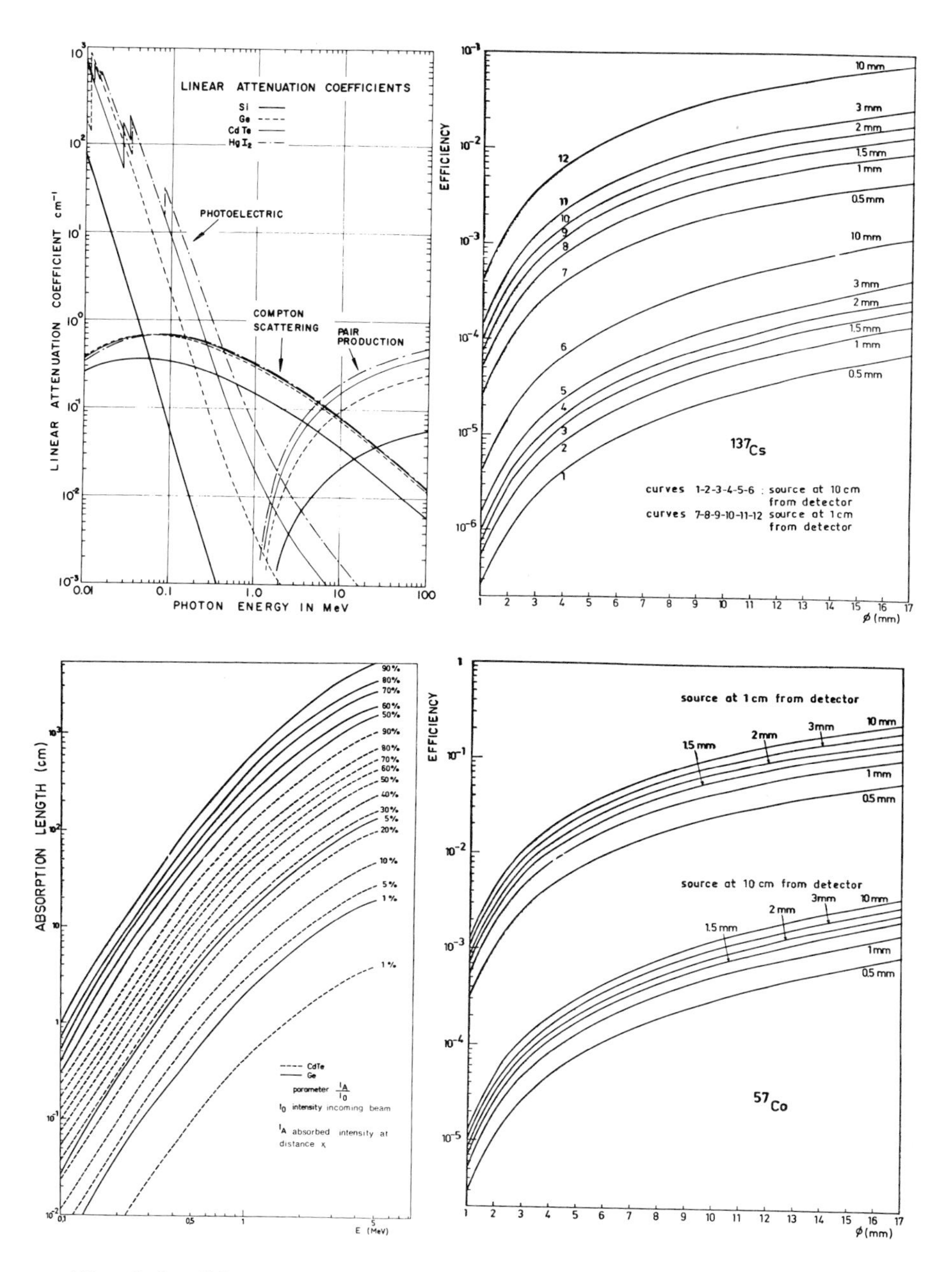

Figs. 8, 9and10:calculated attenuation coefficients and absorption length, as well as various efficiencies for conditions indicated on figures.

where d is the detector thickness, γ_n and γ_p the pathlength of electrons and holes (62, 63).

Similar calculations have been performed for a non uniform field distribution.

Detection efficiency. In nuclear spectroscopy, only counts in the full energy peak are of real interest ($E_\gamma < 1.5$ MeV). Both photoelectric absorption and multiple Compton scattering contribute to this peak. When compared to germanium, the gain in photoelectric efficiency is about a factor 5-6 (at 500 keV) (Fig. 8), whereas, the multiple Compton contribution to the full energy peak can be quite significant, even more than in Ge (63). On figures 9 and 10, we plotted the variation of the calculated relative photoelectric efficiency for two energies (122 and 660 keV) and for two geometries as a function of detector's size. These figures confirm that for the present day available depletion layer thickness the gain in area will bring the highest efficiency increase.

Real situation : To evaluate the real efficiency of the spectrometers, several kinds of measurements are possible :

• determine the relative full energy peak efficiency in comparison to a NaI (Tl) scintillator, for a given detector-source distance. Table V gives some results.

• measure the absolute full energy peak efficiency by using calibrated sources. A typical result, for a small encapsulated spectrometer is reported on Fig. 11.

• use the escape probability under standard conditions as a method to evaluate the real depletion width, or the real resistivity ρ'. Indeed, due to the rather high atomic number of the constituants of the detector, the generated X-rays are of rather high energy (23-27 keV), the escape probability is important (66), varying with depletion layer width X. By simply measuring the escape-to-full energy peak amplitude, it is possible through the calibration of Fig. 12 to evaluate the resistivity ρ' and then calculate depletion width and efficiency, assuming that all the pulses are collected fully, which is true for the spectrometers used in this procedure in which escape peak must be separated from full energy peak (good resolution). Furthermore, the depletion X must be rather small.

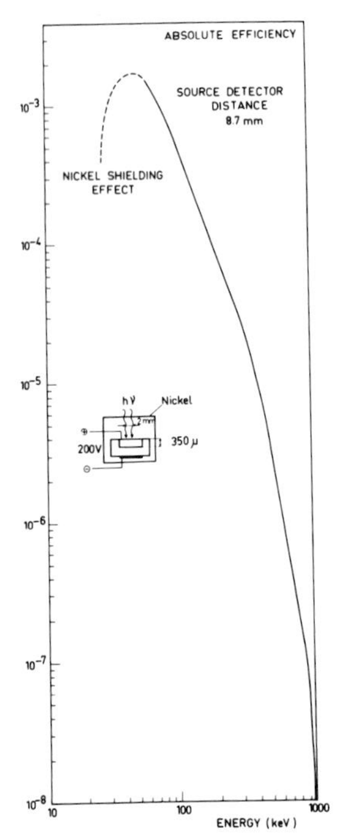

Fig. 11. Measured absolute efficiency as a function of γ-rays energy for a 350 μm thick detector encapsulated as shown on figure.

TABLE V
Full energy peak efficiency of various CdTe detectors relative to a Na(Tl) 1.5 x 1.5" scintillator for a source-detector distance of 10 cm (the thicknesses indicated are that of the whole detector).

DETECTOR	SOURCE	
1. lapped surfaces, with Aluminium contacts.	^{57}Co (122keV)	^{137}Cs (661keV)
diameter 2 mm thickness 0.37mm	0.04 %	0.001 %
diameter 2 mm thickness 0.42mm	0.05 %	0.003 %
diameter 4 mm thickness 2 mm	1.1 %	0.1 %
diameter 4mm thickness 1.35 mm	0.85 %	0.16 %
diameter 6 mm thickness 1.5 mm	2.7 %	0.2 %
2. etched surfaces, with electroless gold contacts surface 30 mm^2 thickness 0.8 mm	0.45 %	0.1 %
3. association of several detectors, with lapped surfaces + aluminium. total surface: 35 mm^2 thickness 1 mm	1.5 %	0.25 %

Loss of efficiency : several contributions to loss of efficiency may exist:
- X-ray escape (we just considered);
- poor charge collection : for a given material of resistivity ρ, the field ξ existing within the depletion zone depends on the applied voltage V, the thickness X but also on the surface handling. For a long time, small $\mu\tau\xi$ values (especially with electroless gold contacts) limited the performance. However, during the last two years sufficient progress has been achieved in crystal growth, to be able to select "good" materials, at least for depletion width values up to 2.0 mm.

Two phenomena can lead to poor charge collection at room temperature : first, long risetime pulses appear, which have not reached their amplitude during the clipping time of the amplifier; this results essentially from a hole trap located at 0.38 eV above the valence band. Then, carriers can also be captured by

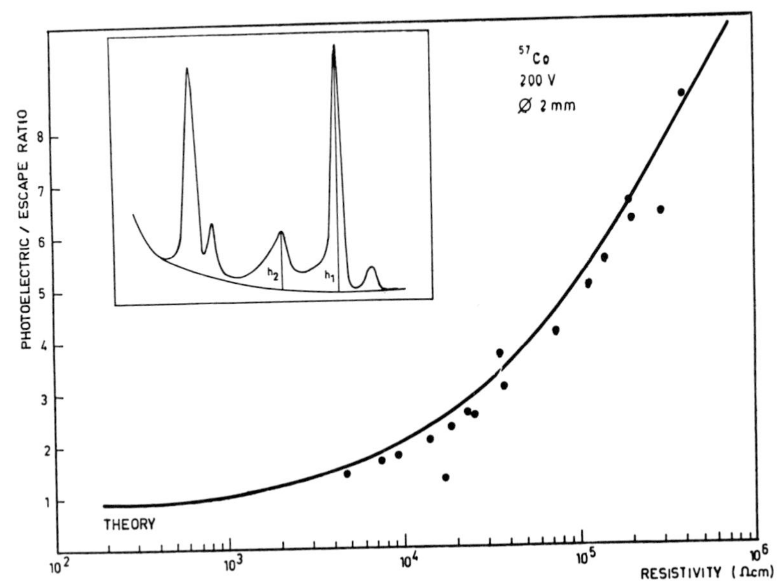

Fig. 12. Photoelectric to escape peak ratio as a function of effective resistivity for a 200 V bias voltage appled. Full line : as calculated; cross : experimental.

deep traps for more or less long time, depending on the particular properties of the deep level (Fig. 13) ;
- polarization : this point will be considered in the section devoted to counter stability.

Energy resolution. Spectroscopists working with Si or Ge detectors are used to having energy resolutions essentially determined by statistical fluctuations in the electron-hole pair generation. Exceptions occur only for very low energy photons, where noise plays a role and charged particles where nuclear energy loss fluctuations contribute to peak broadening. But in both Si and Ge counters, both employed at 77°K in photon spectroscopy, the thermal noise becomes small. Here, the goal being to operate at room temperature, this contribution is no longer negligible. The various contributions to peak broadening are :

Amplifier and detector noise : Diode leakage current and noise of the amplifier produce line broadening which can be calculated. For equal integration and differentiation time constants the results are reported in Fig. 14.

When aluminium is used on small area lapped surfaces, values around 1 keV are observed, they become much larger for electroless gold contact devices. As long as the current is kept in the 10^{-8} to 10^{-9} A range the contribution of this effect is not predominant.

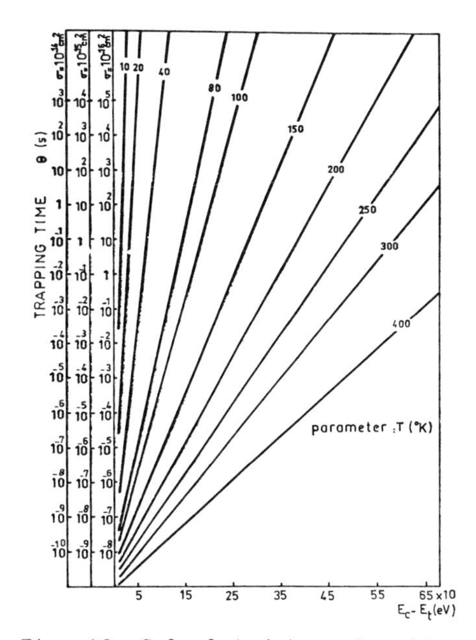

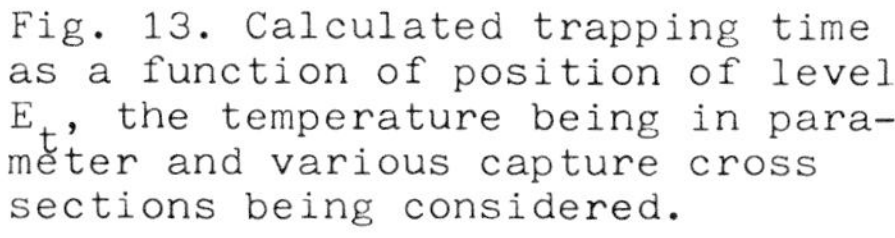

Fig. 13. Calculated trapping time as a function of position of level E_t, the temperature being in parameter and various capture cross sections being considered.

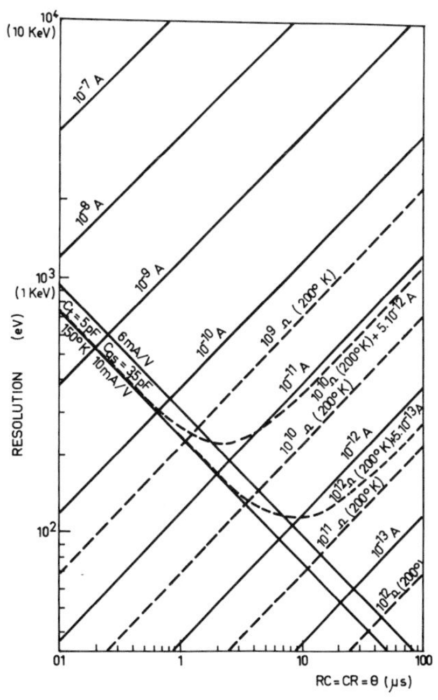

Fig. 14. Calculated FWHM versus shaping time constants.

Ballistic defects : Due to the rather low drift velocity of the carriers, ballistic defects (65) occur for thick detectors. Within the value considered above (2 mm), this effect is no longer determinant.

Trapping-Detrapping : This constitutes the most important contribution to peak broadening. Several authors have considered this effect on peak shape , their models predict either just a gaussian broadening or a tailing on the low energy side of the peak. For a single level, uniformly distributed through the detector, when one type of carrier has a diffusion length much greater than that of the other and also in excess of X, the FWHM L_t is given by :

$$L_t = K\,(1 - \eta)\,E_o,$$

where K is a constant, η the pulse height defect of a photon of energy E_o.

In the case of Ge (Li) spectrometers, the constant K was found to be approximatively the same for all detectors, indicating that the trapping distribution is uniform through the sensitive region. However, it was experimentally observed that L_t increases

with $E_o^{1/2}$ rather than with E_o. In CdTe the situation is as follows for high resistivity material with lapped contacts :
• for a given detector at a certain energy E_o, the above relation is followed : the peak broadening is proportional to pulse height defect η. However, for various counters, the value of K changes from detector to detector, which is an indication of non uniformity and non constant trap distribution.
• when the photons energy is changed, the FWHM degrades as reported on Fig. 15 for a small and a larger detector. The following conclusions can be drawn from these results :
+ at very low energy the resolution is limited by detector noise level ;
+ when energy increases incomplete charge collection is the most pronounced contribution, but the resolution vs energy varies from detector to detector. However, the degradation in resolution vs energy is much faster for the smaller detectors due to escape of photoelectrons out of the sensitive volume.
Some recently recorded spectra obtained with chlorine compensated THM grown materials are given on Fig. 16 and some resolutions on Table VI. For electroless gold or platinum contacts, due to the small electric field, $1 - \eta$ becomes very important and the resolution is strongly degraded; multiple peaking appears and very short amplifier time constants must be used due to the high leakage current. This kind of detector does not constitute today a competitive spectrometer and is much better suited for counting applications without energy resolution.

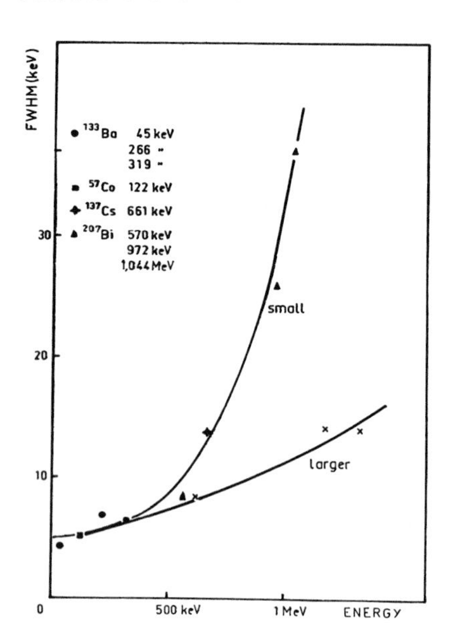

Fig. 15. Resolution (FWHM) vs. γ-ray energy.

We have not considered in this section the possibility to employ CdTe in charged particle spectroscopy. Some experiments have been performed (66, 67) indicating that the resolution is not as good as for Si spectrometers, but it appears that this high density detector can have applications in the high energy physics field (Table VII).

Stability. Short term evolution : "Polarization effect" : up to recently, all crystals exceeding some critical resistivity value ($\rho \simeq 7.10^7\ \Omega$.cm or $\rho' = 7.10^4\ \Omega$ cm), exhibited a progressive reduction of both signal amplitude and counting rate as a function of time the detector was biased. Several models have been proposed as already indicated to explain the effect. They consider either the trapping of detrapping of carriers on a deep level as a function of time. This level is located at about mid-gap, at E_v + 0.7 eV($\pm$ 0.1 eV), has a capture cross section of

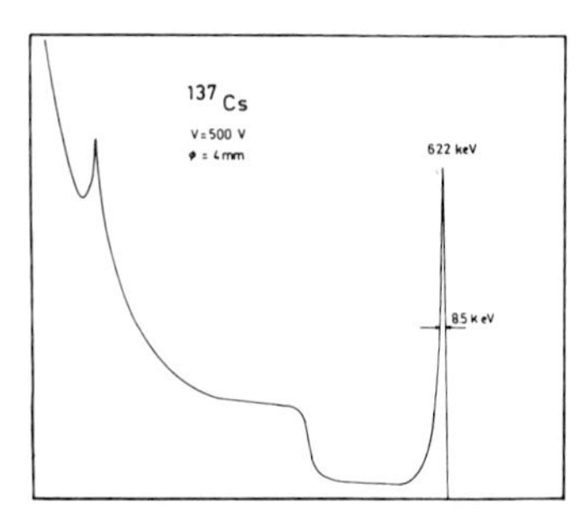

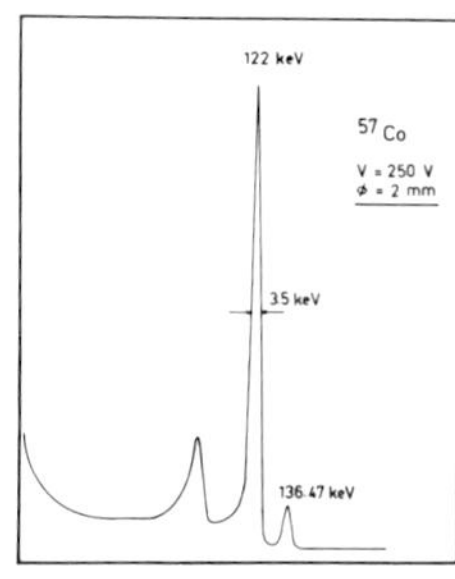

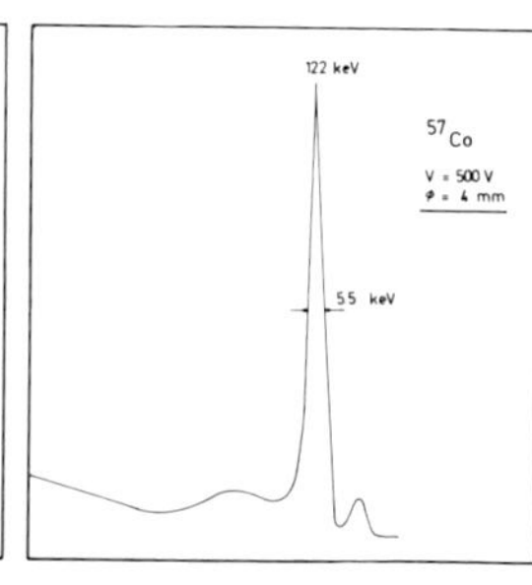

Fig. 16. γ - ray spectra of ^{137}Cs and ^{57}Co recorded with CdTe detectors prepared on semi-insulating chlorine compensated material with lapped surfaces and aluminium contacts. The spectra have been obtained at room temperature.

TABLE VI
Best published energy resolutions obtained on various CdTe

ENERGY (keV)		RESOLUTION (keV)		
		Active volume: 0.3mm^3	Active volume: 5-20mm^3	Active volume 350-400 mm^3
6	^{55}Fe	1.1		
27	^{125}I			9.2
60	^{241}Am	1.7	3.2	12
122	^{57}Co	the active volume is too small, when compared to the photoelectron range.	3.5	
661	^{137}Cs		8	40
1117	^{60}Co		14	

$\sigma = 10^{-16}-10^{-17}$cm^2 and a concentration N_T varying from 10^{10} to 10^{14} cm^{-3}. As we discussed elsewhere (68) its origin is probably related to the presence of doubly charged cadmium vacancies $[V_{Cd}]^{--}$. The mechanism of polarization is as follow :

When the bias voltage V is applied on the detector, the original depletion layer width is expressed by :

$$X^2 = \frac{2\varepsilon V}{q(N_A - N_D)}$$

where all symbols have their normal meaning.

Due to hole emission, a progressive variation of ionized centers occurs with time : $N_A - N_D + N_T(t)$. Therefore, the depletion width decreases with time. If all the centers are uniformly distributed, it becomes :

Detector material:	CdTe (Cl)	Si	Ge(Li)	NaI (Tl)	NE102
Type:	p-type surface barrier diode	n-type surface barrier diode	p-type diode	inorganic scintillator	organic, p scintillato
Operating temp. (K)	250–300	77–300	77	300	300
Density (g/cm^3)	6.06	2.33	5.32	3.67	1.032
E_p (lmm)	16	12	16	13	9
$\bar{Z}$	50	14	32	32	3.5
Diam. x depth (mm)	8 x 1 [g]	20 x 2	50 x 50	50 x 50	50 x 50
Window (mg/cm^2)	0.1–1	0.1–1	>1	>1	0.1–1
Operating voltage	60–100 V	500–1000 V	1–3 kV	1–3 kV [h]	1–3 kV [i]
Bandgap (eV)	1.47	1.12	0.66	n.a.	n.a.
ϵ	4.43 eV	3.61 eV	2.98 eV	2–15 keV [j]	5–50 keV
τ (ns) [k]	100	70	30	230	2.4
Linearity	linear [g]	linear	linear	non-linear	non-linear
$\Delta E/E$ (Z = 1, 10 MeV/amu)	1–2% [g]	0.3–1%	0.3–2%	2–3%	4–6%
$\Delta E/E$ (Z = 8, 5 MeV/amu)	2–3% [m]	0.3–1%	1–2%	3–5%	5–7%
i/e^- (Z = 1, 10 MeV/amu)	≥95%	≥98%	≥98%	70%	50%
i/e^- (Z = 8, 5 MeV/amu)	≥80%	≥98%	≥98%	15%	5%
Allowable dose (n/cm^2)	3×10^{10}	10^{12}	10^{11}	$>10^{10}$	$>10^{10}$

TABLE VII
Characteristics of some conventional charged particle detectors compared to CdTe (67).

$$\frac{X^2(0) - X^2(t)}{(X(t) - \alpha X(0))^2} = \frac{N_T}{N_A - N_D}\,[1-\exp(-e_p t)]$$

where : $\alpha = (E_F - E_T/qV)^{1/2}$

Here $\alpha \ll 1$, therefore :

$$\frac{X^2(0)}{X^2(t)} - 1 \simeq \frac{N_T}{N_A - N_D}\,\{1 - \exp(-e_p t)\} \quad ,$$

where e_p is the deep level emission coefficient given by :

$$e_p = \sigma\, v_{th}\, N_V \exp\left(-\frac{E_T - E_V}{KT}\right),$$

in which v_{th} is the carrier thermal velocity and N_V the number of states in the valence band.

This model is in good agreement with experimental results, even in the case of very severe polarization (Fig. 17).

To overcome this problem, four possibilities exist :

• use lower resistivity material in order to keep the Fermi level away from the deep level E_T so that, when the band is bent, E_T does not cross E_F. This approach has been used in the past for THM "pure" crystal growth ;

- a similar result is achieved by shining light on the contact electrode, increasing the free carrier concentration ;
- reduce the band bending in such a way that E_T cannot cross E_F, in other words apply ohmic contacts and use low electric fields. Such a method has been developed by BELL and WALD (69) with electroless gold or platinum contacts and is now largely employed ;
- suppress the polarization level E_T by a correct choice of the crystal growth conditions. This result was recently achieved in our laboratory even on lapped surfaces with aluminium electrodes. This opens new possibilities to the high resistivity CdTe crystals.

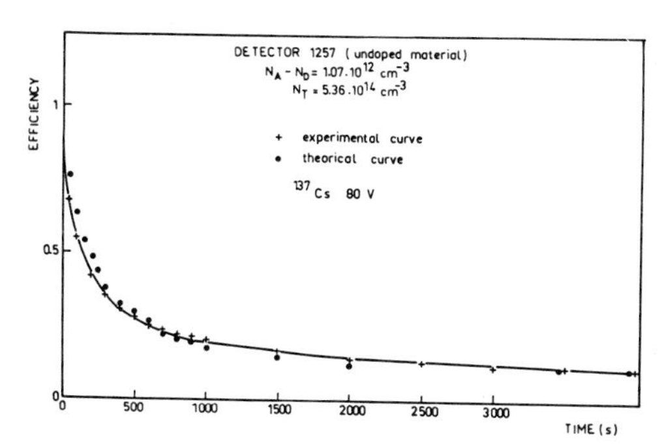

Fig. 17. Evolution of efficiency as a function of time. Full line is calculated and points are experimental values.

<u>Long term evolution</u> :

- Counters prepared by lapping and polishing the surface, which have received aluminium contacts, show no evolution over several years.
- Detectors realized with etched surfaces and electroless gold contacts can, sometimes, give rise to some problems, depending on the preparation procedure, since the etched surface can exhibit evolution with time.

<u>Irradiation effects</u> :

- Polarization can be produced when very high electron-hole pair generation is used, as it has been shown.
- Irradiation of both chlorine and indium doped materials by 33 MeV protons has been considered, as well as strong γ-ray, electron or neutron bombardment (49, 66, 67, 70, 71). For example with ^{60}Co γ-rays an improvement is first observed followed by degradation for fluxes higher than 10^5 rad./h. Annealing at 200°C restored the original performance.
- Damage resulting from ion implantation has also been investigated (73, 74) especially a detailed study of In implantation annealing (75).

PERFORMANCE COMPARISON WITH MERCURIC IODIDE

Since CdTe constitutes, with HgI_2, the only room temperature operating semiconductor γ-ray detector, it is interesting to compare their performance. The theoretical characteristics have been considered by various authors and experimental comparison has recently been performed. We have summarized the major fundamental and experimental characteristics on a series of Tables VIII - IX. It appears that the higher Z of HgI_2 gives no real advantage, since the very low transport velocity of holes limits the sensitive thickness even with a very high field within the detector (up to 5.1 x 10^4 V/cm). But, due to its higher bandgap, which allows the detector to run with currents in the pA range, and to the small energy required per electron-hole with respect to its bandgap, the performance of HgI_2 spectrometers are definitely better for low energy X-ray spectrometry. On other side, the long stability of the HgI_2 devices must be demonstrated and the crystal growth is by far not fully under control.

TABLE VIII
Comparison fundamental properties of CdTe and HgI_2

	CdTe	HgI_2
Structure	cubic (zincblende)	tetragonal
Unit cell (Å)	6.47	a= 4.37 b= 12.44
Density (g/cm^3)	6.06	6.40
Atomic number Z	48 and 52	53 and 80
Gap (ev)	1.45	2.15
Energy per electron-hole (eV) (300 K)	4.43	4.15
Electron mobility (cm^2/V.s.) (300K)	1,000	100
Hole mobility (cm^2/V.s.) (300K)	80-100	4
Saturation carrier velocity (cm/s.) e at a field of	1.5 . 10^7 (11,5 kV/cm)	6. 10^6 (80 kV/cm)
Electron mobility. lifetime product	1.5-3. 10^{-3}	10^{-4}
Hole mobility . lifetime product (cm^2/s.)	3. 10^{-4}	10^{-5}
Schubweg (mm) at field corresponding to maximum velocity (electrons)	150 in practice: 1-2	80 in practice : 1-2
Photoelectric absorption (%) at energy of (for a thickness of 0.5mm)		
10 kev	100	100
20	100	100
50	98	95
100	49	68
200	8	15
500	0.7	1.5
1000	0.15	0.35

APPLICATIONS OF CADMIUM TELLURIDE DETECTORS

Several classifications of the various applications can be considered. Here, we divide the subject into devices without (or low) energy resolution and into spectrometric devices. These detectors have not received up to today the large applications they should, in principle, have essentially for the following reasons :

- the presence of polarization in all of the high resistivity materials limited strongly the stability of the detector. As we have shown above this problem is now solved.
- up to recently, the growth process was not fully under control, as a consequence crystal selection was necessary, leading to a cost increase. In our opinion the selling price of a spectrometer should become the same as that of a silicon surface barrier detector and that of a counter without energy resolution should be near to that of a silicon power rectifier.

- the presence of deep traps leads to a dependency of pulse height on the rate of preirradiation and to photomemory effects i.e. long delay time of signals after end of pulsed irradiation. Since large differences in behaviour appear from crystal to crystal, it will probably be possible to compensate this effect to a degree.

TABLE IX
Performance comparison between CdTe and HgI_2 detectors

	CdTe	HgI_2
Sensitive thickness (maximum) - for spectrometers - for counters	 2mm 6mm	 0.5 - 1 mm a few mm
Active area - for spectrometers - for counters	Al contact: 4mm in diameter Au_2 electroless cm^2.	up to cm^2
Typical bias voltage	100-1000 V	2500 V
can the bias be applied quickly?	yes	no, several hours
Lower energy threshold	3.7 keV	1.25keV (but cooled FET)
Eest reported energy resolution at room temperature and for energies of:		
1.25 keV (X Mg)	noise	385 eV
5.9 keV (^{55}Fe)	1.1 keV	650 eV
60 keV (^{241}Am)	1.7 keV	1.2 keV
122keV (^{57}Co)	3.5 keV	2.5 keV
661keV (^{137}Cs)	8.0 keV	4.5 keV
1.33MeV (^{60}Co)	14 keV	22 keV
Stability under bias (polarization)	up to recently yes, now solved	sometimes, not fully controlled
Degradation with time	no	fragile material some problems
Collimation of active area necessary	no	yes
Crystal growth	difficult, but automatic	difficult, no automatic control is sufficient
Yield of good crystals	high (50 %)	very low (5% ?

<u>Counters without or with poor energy resolution</u>

a) "Safeguard": Due to the small volume and high efficiency, CdTe should be well adapted for individual dosimeters (76, 77) (Fig. 18). Its operation is possible even in the photovoltaïc mode, without external biasing, which is a further advantage over gas (Geiger...) counters.

b) Thickness gauging: By measuring the attenuation of X or γ -rays is largely employed in industry. Cadmium telluride is, in principle, a very good candidate for a high resolution in space systems. It has been used for spacecraft propellant gauging (78) as well as for study of rocket nose cone oblation during the

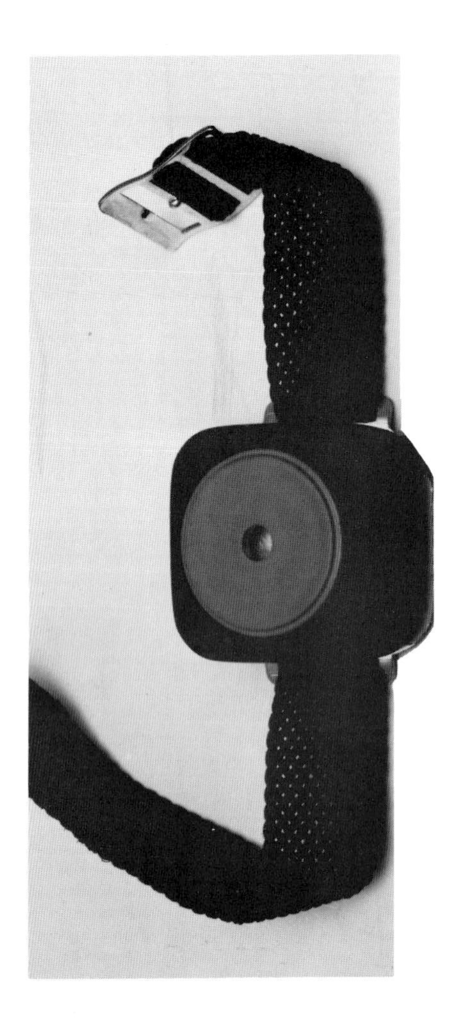

Fig 18: watch shaped safeguard set-up

Fig. 19: Bore hole's wall analysis set-up by X-ray fluorescence.

reentry into atmosphere (79). For these space applications special packaging is needed to prevent noise due to microphony (80). The latter can become a severe problem even on portable set-ups if sufficient care is not given.

c) Nuclear medicine: Since the important work of KAUFMAN et al and ENTINE and BOJSEN, many applications have been considered, some have recently been reviewed (81, 82). Imaging of organs by nuclear techniques is a major objective in this field. Besides the Anger camera which needs a rather good energy resolution, three-dimensional imaging has been developed during the last years ("scanners") using computer reconstruction of the absorption of X-rays seen by an array of detectors. Detectors are operated in the integrated current mode. Since the absorption of the X-rays can change over several orders of magnitude from one measuring point to the next, the photomemory must be negligible. The detectors investigated up to now have all shown some tailing with a decay time constant on the order of 10^{-6} s.

Spectrometers

Both scientific and industrial fields offer applications to CdTe spectrometers, especially, when room temperature, small size, low voltage operating devices with reasonable energy resolution constitute some advantage. Up to now, most efforts have been devoted towards the nuclear medicine field. However, in addition, at least two large possibilities exist in the field of analysis :

Nuclear fuel characterization. The analysis of Plutonium is of particular interest and the peaks at 413,375,345 and 332.3 keV cannot be nicely resolved by a conventional Na(Tl). Therefore, up to now cooled Ge spectrometers have been employed. However, these diodes are rather impractical for a portable system. It was shown in 1973 (83) that a small CdTe spectrometer would fullfill the requirements.

Another important parameter in reactor science is the measurement of the enrichment: $\frac{^{235}U}{^{238}U}$. Due to the small detector size the measurement is possible even on a complete fuel rod assembly.

Finally, CdTe spectrometers have been used to detect failures in heat exchangers in reactors. The spectrometer was inserted between the rows of tubes in each bank of the heat exchanger with nominal spacing as small as 16 mm in the evaporator. Rise-time selection allowed an improvement of energy resolution by rejecting all long rise-time pulses.

Chemical analysis in bore holes. Minerals, including uranium, as well as oil can be detected directly on the wall of a bore hole either by fluorescence X-ray excitation or by (n,γ) reactions. The technique has been developed using NaI(Tl) as a detector, but the limited resolution gives rise to problems. Ge spectrometers have been tested but the cooling in the severe conditions is difficult and very limited in time. CdTe has shown promising capabilities. A system recently developped is shown on figure 19.

CONCLUSION

During the last two years CdTe detectors have been noticeably improved since high resistivity crystals of very large size can be grown, which are free of polarization. The depletion layer width, of about 2 mm is sufficient for most of the current applications. Rather large area counters with poor energy resolution can now be produced for safeguard or medical applications ; for spectrometers having good energy resolution the active diameters are still limited by excess noise for large areas, this problem needs further investigation. The progress on these spectrometers has not been in the last ten years as fast as expected. This results, essentially, from the complexity of this material when compared to Si or Ge, but also from the very limited number of people involved in this subject.

REFERENCES

1. G.A. ARMANTROUT, S.P. SWIERKOWSKI, J.W. SHEROHMAN, J.H. YEE, IEEE Trans. Nucl. Sci. NS-24 (1977) 121.
2. G.A. ARMANTROUT, Nucl. Instr. Meth. 193 (1982) 41.
3. G.L. MILLER, IEEE Trans. Nucl. Sci. NS-19 (1972) 251.
4. B. BONAND, A. FRIANT, Report CEA-BIB-210 (1974).
5. Papers on 2nd Intern. Symp. on Cadmium Telluride Strasbourg (1976) published in Rev. de Phys. Appl. 12 (1977) n°2.
6. Intern. Workshop on mercuric iodide and cadmium telluride nuclear detectors (1977) JERUSALEM, published in Nucl. Instr. Meth. 150 (1978).
7. P. SIFFERT, Rev. Phys. Appl. C3, 39 (1978) 40.
8. K. MULLER, K. PROKERT, N. EITEL JORGE, K. SCHILLINGER, Kernenergie 22 (1979) 41.
9. R.C. WHITED, M.M. SCHIEBER, Nucl. Instr. Meth. 162 (1979) 113.
10. E. SAKAI, Intern. Symposium on Nucl. Rad. Detectors TOKYO (1981).
11. W.R. HARDRING Nature 187 (1960) 405.
12. P. SCHUSTER, N. GELOFF Acta Phys. Austr. 23 (1966) 387.
13. S. ASIMOV, Atom. Energ. Sov. 40 (1976) 346.
14. J.E. EBERHARDT, R.D. RYAN, A.J. TAVENDALE, Appl. Phys. Lett. 17 (1970) 427.
15. J.E. EBERHARDT, R.D. RYAN, A.J. TAVENDALE, Nucl. Instr. Meth. 94 (1971) 463.
16. P.E. GIBBONS, J.H. HOWES, IEEE Trans. Nucl. Sci. NS19 (1972) 353.

17. K. HESSE, W. GRAMANN, D. HOPPER, Nucl. Instr. Meth. 101, (1972) 39.

18. T. KOBAYASHI, T. SUGITA, M. KOYAMA, S. TAKAYANAGI, IEEE Trans. Nucl. Sci. (1972) 324, id. NS 20 (1973), 318, NS 23 (1976) 97.

19. K. ZANIO, Semiconductors and semimet. 13 (1978) Eds. RK WILLARDSON and A.C. BEER, Academic Press.

20. R.O. BELL, F. WALD, C. CANALI, F. NAVA, G. OTTAVIANI, IEEE Trans. Nucl. Sci. NS 21 (1974) 331.

21. F.A. KROGER, J. Phys. Chem. Solids 26 (1965) 1717.

22. R. STUCK, A. CORNET, C. SCHARAGER, P. SIFFERT, J. Phys. Solids 37 (1976) 989.

23. R.N. BHARGAVA, J. Crystal Growth 59 (1982) 15.

24. E. MOLVA, J.P. CHAMONAC, J.L. PAUTRAT, Phys. Stat. Sol. (b) 109 (1982) 635.

25. P. HOSCHL, P. POCIVKA, V. PROSSER, M. VANECEK, Rev. Phys. Appl. 12 (1977) 229.

26. E.N. ARKADEVA, O.A. MATVEEV, E.V. MELNIKOVA, A.I. TERENTEV, Sov. Phys. Semicond. 14 (1980) 839.

27. B. SCHAUB, provate communication.

28. E.N. ARKADEVA, O.A. MATVEEV, E.V. MELNIKOVA, Sov. Phys. Semicond. 14 (1980) 424.

29. A.J. DABROWSKI, J. IWANCZIK, R. TRIBOULET, Nucl. Instr. Meth. 126 (1975) 417.

30. A.T. AKOBIROVA, L.V. MASLOVA, O.A. MATVEEV, A.K. KHUSAINOV, Sov. Phys. Semicond. 8 (1975) 1103.

31. M. HAGE-ALI, R. STUCK, A.N. SAXENA, P. SIFFERT, Appl. Phys. 19, (1979) 25.

32. M.H. PATTERSON, R.H. WILLIAMS, J. Phys. D (Appl. Phys. 11) (1978) L 83.

33. C.A. MEAD, W.G. SPITZER, Phys. Rev. 134 (1964) 935.

34. A.T. AKOBIROVA, L.V. MASLOVA, O.A. MATVEEV, A.K. KHUSAINOV, Sov. Phys. Semicond. 8 (1975) 1103.

35. R. SWANK, Phys. Rev. 153 (1967) 844.

36. J. TOUSKOVA, R. ZUZEL, Phys. Stat. Sol. (a) 15 (1973) 257 id. (a) 40 (1977) 309.

37. M.H. PATTERSON, R.H. WILLIAMS, J. Cryst. Growth 59 (1982) 281.

38. M. SARAPHY, Thesis Strasbourg, 1980.

39. F.A. KROEGER, D. DE NOBEL, J. Electron. 1 (1955) 290.

40. M. HAGE-ALI, R. STUCK, A.N. SAXENA, P. SIFFERT, Appl. Phys. 19 (1979) 25.

41. M.H. PATTERSON, R.H. WILLIAMS, J. Appl. Phys. D (Applied Physics) 11 (1978) 283 , J. Crystal Growth 59 (1982) 281.

42. A.T. AKOBIROVA, L.V. MASLOVA, O.A. MATVEEV, A.K. KHUSAINOV, Sov. Phys. Semicond. 8 (1975) 1103.

43. F.W. WALD, Rev. Phys. Appl. 12 (1977) 277.

44. M. CHU, A.L. FAHRENBRUCK, R.H. BUBE, J.F. GIBBONS, J. Appl. Phys. 49 (1978) 322.

45. J.C. BEAN, Thesis Stanford University (1976).

46. C.B. NORRIS, C.I. WESTMARK, G. ENTINE, S.A. LIS, H.B. SERREZE Rad. Effects Lett. 53 (1981) 115.

47. C. AN, H. TEWS, G. COHEN, J. Cryst. Growth 59 (1982) 289.

48. H.L. MALM, M. MARTINI, Con. J. Phys. 51 (1973) 2336 ; id. IEEE Trans. Nucl. Sci. NS 21 (1974) 322.

49. P. SIFFERT, R. BERGER, C. SCHARAGER, A. CORNET, R. STUCK, R.O. BELL, IEEE Trans. Nucl. Sci. NS 23, (1976) 159.

50. R.O. BELL, G. ENTINE, H.B. SERREZE, Nucl. Instr. Meth. 117 (1974) 267.

51. M. HAGE-ALI, C. SCHARAGER, J.M. KOEBEL, P. SIFFERT, Nucl. Instr. Meth. 176 (1980) 499.

52. R.J. FOX, D.C. AGOURIDIS, Nucl. Instr. Meth. 157 (1978) 65.

53. G. ENTINE, M.R. SQUILLANTE, H.B. SERREZE, IEEE Trans. Nucl. Sci. NS 28 (1981) 558.

54. N.V. AGRINSKAYA, O.A. MATVEEV, Sov. Phys. Semicond. 14 (1980) 611.

55. P.G. KASHERMINOV, O.A. MATVEEV, L.M. MASLOVA, Sov. Phys. Semicond. 3 (1969) 451.

56. A. CORNET, P. SIFFERT, A. COCHE, R. TRIBOULET, Appl. Phys. Lett. 17 (1970) 432.

57. B. RABIN, H. TABATABAI, P. SIFFERT, Phys. Stat. Sol. (a) 49 (1978) 577.

58. P. SIFFERT, B. RABIN, H. TABATABAI, R. STUCK, Nucl. Instr. Meth. 150 (1978) 31.

59. P. SIFFERT, Nucl. Instr. 150 (1978) 1.

60. S.I. ZATOLOKA, V.P. KARPENKO, P.G. KASHGRININOV, O.A. MATVEEV, D.G. MATYUKHIN, A. ATOMASOV, V.S. KHRUNOV, Sov. Phys. Semicond. 13 (1979) 981.

61. O.G. KASHERININOV, O.A. MATVEEV, D.G. MATYUKHIN, Sov. Phys. Semicond. 13 (1979) 756.

62. R.B. DAY, G. DEARNALEY, J.M. PALMS, IEEE Trans. Nucl. Sci. NS 14 (1967) 487.

63. P. SIFFERT, J.A. GONIDEC, A. CORNET, R.O. BELL, F. WALD, Nucl. Instr. Meth. 115 (1974) 13.

64. H. JAGER, R. THIEL, Rev. Phys. Appl. 12 (1977) 293.

65. P. SIFFERT, A. CORNET, R. STUCK, R. TRIBOULET, Y. MARFAING, Trans. Nucl. Sci. NS 22 (1975) 211.

66. G.H. NAKANO, W.L. IMHOF, J.R. KILNER, IEEE Trans. Nucl. Sci. NS 23 (1976) 468.

67. R.A. RISTINEN, R.J. PETTERSON, J.J. HAMILL, F.D. BECCHETTI, G. ENTINE, Nucl. Instr. Meth. 188 (1981) 445.

68. P. SIFFERT, M. HAGE-ALI, R. STUCK, Rev. Phys. Appl. 12 (1977) 335.

69. F.V. WALD, R.O. BELL, Contract report AT - (11-1)-3545 (1975).

70. C.B. NORRIS, C.E. BARNES, K. ZANIO, J. Appl. Phys. 48 (1977) 1659.

71. C.E. BARNES, C. KIKUCHI, Rad. Eff. 2 (1970) 243.

72. T. TAGUCHI, Y. INUISHI, J. Appl. Phys. 51 (1980) 4757.

73. M. GETTINGS, K.G. STEPHENS, Rad. Eff. 22 (1974) 53.

74. O. MEYER, E. LANG, Intern. Symp. on Cdte (1971) Ed. P. SIFFERT and A. CORNET.

75. R. KALISH, M. DEICHER, G. SCHATZ, J. Appl. Phys. 53 (1982) 4793.

76. R.J. FOX, D.C. AGOURDIS, Nucl. Instr. Meth. 157 (1978) 65.

77. M.A. WOLF, C.J. UMBARGEW, G. ENTINE, IEEE Trans. Nucl. Sci. NS 26 (1979) 776.

78. F. BUPP, M. NAGEL, W. AKUTAGAWA, K. ZANIO, IEEE Trans. Nucl. Sci. NS 20 (1973) 514.

79. C.R. DROMS, W.R. LANGDON, A.G. ROBINSON, G. ENTINE, IEEE Trans. Nucl. Sci. NS 23 (1976) 498.

80. R.B. LYONS, Rev. Phys. Appl. 12 (1977) 385.

81. J.M. SANDERS, N.M. SPYROU, Nucl. Instr. Meth. 193 (1982) 79.

82. P.A. GLASOW, B. CONRAD, K. KILLIG, W. LICHTENBERG, IEEE Trans. Nucl. Sci. NS 28 (1981) 563.

83. W. HIGINBOTHAM, K. ZANIO, W. AKUTAGAWA, IEEE Trans. Nucl. Sci. NS 20 (1973) 510.

DETECTOR FABRICATION AND MATERIAL STUDIES

$Bi_4Ge_3O_{12}$ (BGO) - A Scintillator Replacement for NaI(Tl)

M. R. Farukhi
Harshaw Chemical Company, 6801 Cochran Rd., Solon, Ohio 44139 USA

ABSTRACT

Scintillation performance of recent (2X) grown BGO is studied. Results indicate BGO to perform better than 8% ^{137}Cs FWHM NaI(Tl) at energies 2.6 Mev and higher. Even the low energy performance of BGO is suitable for considering it in lieu of NaI(Tl) for many spectroscopic applications.

INTRODUCTION

The history of NaI(Tl) as a useful scintillator started with Hofstadter's observation [1] 1948:

> "In tests made by placing crystals of NaI, KI, and naphtalene on photographic plates much greater light output was observed from NaI than from naphtalene samples of comparable size..."
>
> Phys. Rev. 74:100 (1948)

Hofstadter [2] then added a "pinch of thallium halide" to NaI and since 1948, NaI(Tl) became the primary scintillation detector of choice for applications in: [3]

1. Geophysical and environmental science
2. Nuclear fuel cycle
3. Nuclear medicine
4. Space science
5. High energy physics

KI and KI(Tl) were grown first and investigated for scintillation detection but its use was limited due to its inherent radioactivity and somewhat poor light output compared to NaI(Tl). Scintillating alternates such as CsI(Tl), LiI(Eu), CsI(Na) and CaF_2(Eu) made their debuts at various times in the chronological history of inorganic scintillating crystalline phosphors, but each one of these scintillators found their own niche for particular applications requiring their unique characteristics. NaI(Tl) has continued to occupy the mainstay position to date.

Luminescence of BGO to x-ray excitation was first observed and studied by Weber and Monchamp [4] in 1973. A year later, Nestor and Huang [5] presented data on the scintillation response to gamma rays from ^{57}Co, ^{137}Cs, ^{22}Na and ^{241}Am-alpha particles. The scintillation response compared to NaI(Tl) did not elicit anything more than a curious interest for spectroscopic applications. (See Table 1).

TABLE 1
Scintillation Properties of BGO and NaI(Tl)

	Pulse Height Relative Units	^{137}Cs-Resolution % FWHM	Decay Constant nsec	Wave Length Emax nm	Density g/cc
BGO	8	15	300	480	7.13
NaI(Tl)	100	<7	230	415	3.67

The acceptance of BGO for commercial and research applications stemmed from a property (afterglow) that had not been recognized as significantly important to measure. Indeed for spectroscopic applications it is not of primary concern as evidenced by the wide variation in afterglow in NaI(Tl). The x-ray CT instrument manufacturers starting from AS&E and Ohio Nuclear (now Technicare) adopted the use of BGO due to its high stopping power, non-hygroscopic nature and most importantly, a lack of any measurable afterglow even to very high incident x-ray flux. It became the detector of choice from **1975-1978 replacing NaI(Tl)**---the first scintillation detector used by Hounsfield[6] in the discovery of the x-ray CT technique.

Developments 1975 - 1982

Afterglow studies in BGO and other scintillators were conducted in 1977 by Farukhi [7] in collaboration with Cho and Mattson. BGO demonstrated the lowest value for afterglow with $CdWO_4$ having a comparable value. A value of less than 0.005% at 3 ms for x-rays generated in the range of 60 - 150 kev is a commonly accepted value. More exact studies are needed and will most probably be forthcoming as present CT scanners approach the sub-millisecond scan times for medical diagnostic imaging.

$CdWO_4$ has replaced BGO for x-ray CT applications due to its higher light output and emission in the longer wavelengths (540 nm) providing for better spectral match and performance to silicon-diodes. The increasing use of $CdWO_4$ in XCT scanning forebode the end of BGO as a useful scintillator were it not for its possible use in spectroscopic applications. Consequently, problems in material purification and crystal growth perfection were addressed at Harshaw leading to what is colloquially termed "spectroscopic grade" or "twice" (2X) or "thrice" (3X) crystallized $Bi_4Ge_3O_{12}$.

Spectroscopic applications for BGO started with its use in positron CT imaging. Cho and Farukhi [8] reported on the potential use of BGO for positron CT imaging based on coincidence resolving times of 7.0 ns FWHM for 511 kev annihilation gammas and a four-fold photofraction advantage over NaI(Tl). Based on these observations alone, Thompson, Yamamoto and Meyer [9] in collaboration with Farukhi and the technical staff at Harshaw took the bold step in contracting to have the latter build the first BGO positron tomograph for installation at the Montreal Neurological Institute. This machine has taken over 3500 scans to date. Several commercial and research scanners are presently in existence undergoing continuing improvements in imaging and functional parameters.

The impasse with NaI(Tl) was overcome and BGO has replaced NaI(Tl) for positron CT applications and in particular in those systems striving for high efficiency and high resolution. CsF and BaF_2 are viable alternatives to BGO by virtue of their fast decay and coincidence resolving times. These and some other scintillators [10] offer the possibility of using time-of-flight information in the reconstruction to yield a factor of 2 improvement in the image contrast. But none of these alternatives offer the general appeal of BGO for spectroscopic applications.

Table 2 tracks the chronological history of coincidence resolving times (CRT) for BGO in the positron CT field and indicates certain trends:

1. The BGO-BGO resolving times have continued to improve and are directly traceable to the quality of the crystal as reflected by light output which in turn minimizes the jitter. (See Table 3 for light output improvement.)

TABLE 2

Coincidence Resolving Times (511 kev) for BGO with Chronological Improvement in Crystal Quality and Performance

Year	Start/Stop	FWHM ns	Threshold kev	PMT Type	PMT Dia. in	BGO Size cm	Reference
1976	BGO-BGO	7.0	100	RCA 8575	2	2.0x3.8	Cho and Farukhi [8]
1977	BGO-BGO	15-20	350	HTV 1213	0.75	1.8/2.2x3x3	Thompson, Yamamoto and Meyer [9]
1979	BGO-BGO	5.0	300	Ampx. PM 1910	0.75	2.0x3.8	Carroll, Hendry and Currin [11]
1980	BGO-BGO	5.2	350	Ampx. PM 1910	0.75	1.2x2.0x2.6	Nohara, Tanaka, Tomitami et. al. [12]
1981	BGO-BGO	5-10	100	HTV R647	0.5	0.8x2x3	Farukhi (13)
1981	BGO-BGO	3.6	350	HTV R1362	1.125	1.5x2.4x2.4	Murayama, Nohara, Tanaka and Hayashi [14]
1981	BGO-BGO	2.1	350	HTV R1362	1.125	1.5x2.4x2.4	Takami, Ishimatsu, Hayashi et. al [15]
1982	BGO-BGO	2.9	350	HTV R1548	Dual	1.2x2.4x2.4	Yamashita, Ito and Hayashi [16]
1982	BGO-BGO	2.0	$\backsimeq$ 350	HTV R329-2	2	1.5x2.4x2.4	Okajima, Takami, Ueda et. al. [17]
1979	CsF-BGO	3.2	100	Ampx. XP2020	2	2x3.2	Allemand, Gresset and Vacher [18]
1979	CsF-BGO	2.4	250	RCA 8575	2	2.5 Cube	Mullani, Ficke, Ter-Pogossian [19]
1981	CsF-BGO	2.1	100	C 31024	2	2x3.2	Moszynski, Gresset, Vacher et. al. [20]
1982	NE104-BGO	2.3	350	HTV R1548 Rect.	Dual 0.47 x 0.94	1.2x2.4x2.4	Uchida, Yamashita et. al. [21]
1982	NE104-BGO	1.6	350	HTV R1362	1.125	1.5x2.4x2.4	Takami, Ishimatsu, Hayashi et. al. [16]
1982	NE104-BGO	1.4	350	HTV R329-2	2	1.5x2.4x2.4	Okajima, Takami, Ueda et. al. [18]

The material studied by Cho and Farukhi [8] was single crystallized (i.e. grown from powder) and the material tested by Okajima et al [17] is thrice crystallized and reported to have twice the light output of the 1977 material. Though light output comparisons are relative (they are not measured in absolute terms) and quite arbitary, we do confirm the thrice grown material today is a factor of 2 better than the material supplied by Harshaw pre-1980.

2. The CRT of 2.0 ns FWHM for BGO-BGO and 1.4 ns FWHM for plastic-BGO is certainly comparable to NaI(Tl).

3. Various photomultipliers have been used for timing and for light sensing in positron tomographs. These PMTs were not particularly designed for optimium timing performance or for better spectral match to BGO. The effort by Hamamatsu TV in addressing this need is commendable in that the rectangular tube R 1548 is designed specifically for BGO use in positron CT instrumentation, while the R 1362 (not available outside Japan) combines the best in quantum efficiency for BGO emission as well as timing.

4. PMT development is still needed to tailor a response [just as the S-11 was pioneered for NaI(Tl)] for the BGO scintillation emission spectra.

Moszynski et al [20] have studied the pulse shape of BGO on Harshaw supplied material that was crystallized (1.5X) and 2 cm dia. x 3 cm in. thickness. Their study revealed several interesting and exciting prospects for the future:

1. The rise time of BGO is 2.8 ns for the sample dimension studied and is comparable if not superior to NaI(Tl). Further studies by the same group [22] on 1 mm thick NaI and 3 mm thick BGO indicated a very fast rise time of 180 ps for BGO and a much slower rise time for NaI(Tl) of 310 ps (multiple light reflections in the larger area NaI(Tl) piece could deteriorate the rise time).

2. The decay curve for BGO shows an initial fast decay component of 60 ns with 10% the total light and the main component at 300 nsec. Our preliminary investigation confirms this fast component and we observe an <u>additional</u>, longer but weaker component in the range 570-600 ns.

3. NaI(Tl) displays an initial slower decay constant followed by the well-known main component at 230 ns. Moszynski et al [20] indicated that phosphoresence or afterglow in NaI(Tl) precludes the triggering off the first photoelectron and in fact it takes over 3 - 4 photo-electrons to effectively trigger in coincidence measurements. BGO has an absence of afterglow and the initial decay allows one to work with true triggering off the first photoelectron. Hence, Moszynski et al [20] offer the reason why the BGO timing is no worse than twice that of a comparable NaI(Tl) crystal even though the pulse height or light yield is 6%-10% relative to NaI(Tl).

4. This particular characteristic will offer timing advantages for BGO anti-compton shields for Ge-Spectrometers.

<u>Light Conversion Efficiency</u> is often reported as pulse height or integrated charge measurements relative to NaI(Tl) for gamma and x-ray excitation [7].

Pulse height measurements are relative and not absolute measurements since little if any correction or allowance is made for the spectral emission mismatch or the time constant differences for various scintillators. Both factors affect pulse height and besides not all NaI(Tl) sold in the market place displays a constant pulse height. A variation of ± 10% is common and it is not unusual to find 20% differences in crystal samples based on the location and size of the ingot from which they are taken.

A better criterion of measurement is the energy resolution for ^{137}Cs or ^{60}Co; this is the standard way of specifying solid state nuclear detectors used in spectroscopic applications. Table 3 tracks the historical improvement in light conversion efficiency for BGO.

The early material had a yellowish cast and yielded values of 15% FWHM for 3mm thick samples grown from powdered raw material (1X). The (1.5x) material is a mixture of powder and once grown scrap and was referred to as "positron" grade at Harshaw in deference to the market it was serving. The (2x) material of Hitachi has been reported to yield 11.6% FWHM for 662 kev gamma for pieces 15x24x24 mm^3 and 10% FWHM for (3x) grown on R878 PMT. The R1306 is a recent improvement in PMT technology by Hamamatsu yielding PMTs that consistently demonstrate 6.4% - 6.6% FWHM ^{137}Cs resolution for NaI(Tl) standard crystals.

Okajima et al [17] used this PMT to report energy resolutions for Hitachi (3x) material:

Energy (kev)	FWHM	Sample
662	9.5%	15x24x24mm^3
511	10.7%	
1172	7.4%	
1332	6.8%	

Measurements on Harshaw (2x) material show:

Energy (kev)	FWHM	Sample
662 kev	9.3%	25.4mm dia. x 25.4mm
511	10.6%	
1172	7.4%	
1332	6.7%	

This improvement in resolution from 15% FWHM for ^{137}Cs to 9.3% translates into a factor of 2 improvement in the pulse height assuming equal surface preparation, coupling, etc. A more significant observation is that a crystal as large as 51mm dia. x 43mm thick should yield 10.5% FWHM for ^{137}Cs. Upon unpacking this sample the interface (10^6 CSt viscosity silicone oil) was found to be 3mm thick with Al_2O_3 powder leaked into the interface - a less than optimum encapsulation procedure. This crystal will be further studied along with (3x) material of various lengths.

Experimental

Scintillation samples that were (2x) times crystallized were chosen for investigating their response to gamma rays. These samples were deliberately chosen to be larger in volume to samples listed in Table 3:

1. Sample #JH-1-2X 1" dia. x 1" thick
2. Sample #BU-1-2X 1" dia. x 1" thick
3. Sample #32R-055 51mm dia. x 43mm thick

TABLE 3

Improvement in the Scintillation Response of BGO

Year	Crystal Size (mm)	Time Cryst.	PMT Type	PMT Dia.	RPH NaI(Tl)	662 kev % FWHM	511 kev % FWHM	1172 kev % FWHM	1132 kev % FWHM	Reference
1974	25 dia. x 3	1x	RCA 8850	2"	8	15	16			Nestor et. al. [5]
1976	20 dia. x 38	1.5x	RCA 8575	2"	10	15-20	16-21			Cho, Farukhi [8]
1977	2 x 3 x 3	1.5x	RCA 4516	0.75"	--	25-30	26-32			Thompson [9]
1979	1" cube	1.5x	HTV R980	1.5"	11		17.5			Mullani [19]
1979	38 dia. x 38	1.5x	Ampx. XP2000	5"	10	15.4		Not Resolv.	Est. 11.8	A. Evans [23]
1981	15 x 24 x 24	2x	HTV R878	2"	12	11.6		9.3	8.3	Hitachi data Sheet 5/28/82
	15 x 24 x 24	3x	HTV R878	2"	12	10.0		7.7	7.0	Hitachi data Sheet 5/28/82
1982	15 x 24 x 24	3x	HTV 1306	2"	15	9.5	10.7	7.4	6.8	Okajima [17]
1981	1" cube	3x	HTV 1306	2	16					Takami [15]
1982	25 dia. x 25	2x	HTV 1306	2		9.3		7.4	6.7	JH-1-2X
1982	51 dia. x 43	2.5x	HTV 1306	2		1 0.5	11.9	8.4	8.1	32R-055

The end faces were polished and the sides were roughned with #240 Emery paper. One end face was coupled to the PMT face with optical coupling grease G-688 and packed with Al_2O_3. Methylene iodide was also used as a high index coupling fluid but was abandoned as it turned brown.

Signal processing electronics consisted of a Harshaw NB-11 preamplifier, NA-24 linear amplifier set to 2 μs time constant and a Tracor TN1705 MCA with 1024 conversion gain. Several HTV R1306 PMTS were tested and the results are tabulated below.

Sample	PMT S/N	High Voltage	^{137}Cs Res. %FWHM	Low/High Half-Width	^{137}Cs Res.* NaI(Tl)
JH-1-2X	CE4237	700	9.3	37/37	6.3
JH-1-2X	CE454	800	9.5	38/38	6.4
JH-1-2X	CE4195	700	9.3	37/39	6.4
JH-1-2X	CE4509	700	9.7	38/39	6.4
BU-1-2X	CE4237	700	9.6	37/40	6.3
32R-055	CE4237	700	10.5	46/45	6.4
32R-055	R1307**	900	10.7	41/45	6.4**

*Measured at Hamamatsu TV Co (Japan) on NaI(Tl) standard reference crystal (2"dia. x 2" thick) Type 8D8 S/N LW 674 supplied by Harshaw Chemical Company.

**R1307 is a 3" PMT developed initially by Hamamatsu specifically for Harshaw Heavy-ion spin "crystal ball" spectrometer supplied to MPI (Heidelburg).

The number of samples and PMT's tested were to insure that the results reported were not an isolated case and really represent the state-of-art.

Gamma ray spectra for ^{137}Cs and ^{60}Co is displayed is Figure 1 and 2 for sample JH-1-2X. Notice the clear separation of the ^{60}Co peaks. Sample 32R-055 is larger than Evans' [23] crystal but smaller than Drake's [25].

38mm dia. x 38mm < 51mm dia. x 43mm < 76mm dia. x < 76mm
Evans [23] < 32R-055 < Drake [25]

Gamma ray spectra for ^{137}Cs, ^{60}Co, ^{22}Na and ^{228}Th is displayed in Figures 3,4,5 and 6, respectively. Note the significant improvement in resolution for the ^{60}Co peaks when compared to Harshaw sample of Evans [23] ("...not quite good enough to resolve the 1.17- and 1.33 Mev photopeaks").

Evans [23] pioneering work in extending the use of BGO for larger detectors and higher energies compared the performance of a 8% FWHM 662kev NaI(Tl) to an equivalent size BGO detector. The following improvement in the BGO response <u>to date</u> (based on this investigation) may be noted:

1. (9.3% Cs) - BGO yields a ^{60}Co resolution similar to (8% Cs)-NaI(Tl). Table 3 sample JH-1-2X
2. (10.5% Cs) BGO yields a 5.7% resolution for 2.62 Mev compared to 5.8% for 2.75 Mev. (8%Cs) NaI(Tl)
3. A (7.2%Cs) 2" dia. x 2" integral line standard NaI(Tl) unit displayed 5.1% for 2.62 Mev.

Data from these investigations is compared in Figure 7. If a linear projection is made, the sample 32R-055 should yield 2% FWHM for 20 Mev. Even better results can be anticipated for (3x) material and will be the subject of a later investigation.

Discussion:

The growth of crystal from a solution of its constituents is by its very nature a purification process. The impurities have a higher solubility in the liquid and generally remain there while the solid crystal is frozen out. Successive growth of crystals is much akin to zone refining and hence it is not surprising to find that (2x) or (3x) material to be superior in crystal purity to singly grown material. Takagi et al [25] have investigated BGO with respect to successive growth and the addition of impurities to study void formation. Recrystallization of more than three times was found not to yield any appreciable improvements.

This study indicates that raw material purity considerations are not well understood but can be offset by successive growth. This is not a fully acceptable process from an economic viewpoint. Material quality of (2x) will yield small crystals which could match the performance of NaI(Tl) in the region of 2 Mev and up; but larger crystals still have inclusions due to non-stoichiometry and this problem needs to be solved.

Energy resolutions between 9-10% FWHM for ^{137}Cs are certainly interesting enough for BGO to be considered in lieu of NaI(Tl) for many applications involving low energy x-and gamma ray excitation. The 8%-9% range is goal for future improvement in BGO. Recent investigations of BGO [26, 13, 20] have shown that the process is not a simple one mode mechanism. There appear to be 3 decay times and at least 3 emission maxima (480, 530, 570 nm) and possibly a 4th one at 610 nm. Understanding and optimizing these features are the next logical step in the development of BGO.

The timing and photoforaction advantages of BGO for positron annihilation gamma rays have caused it to replace NaI(Tl) in positron CT imaging. Analytical formulae for image forming as well as comparision of BGO to other detector materials in positron tomographs have been studied by Derenzo [27]. The use of BGO in other spectrocopic applications has been initiated and it is replacing NaI(Tl) by virtue of providing cleaner spectra, ease of handling and higher detection efficiency for similar sizes. References are noted below:

Geophysical Exploration

1. Comparision of sodium iodide, cesium iodide and bismuth germanate scintillation detectors for borehole gamma-ray logging.
   D. C. Stromswold. IEEE NS-28, 290 (1981)

2. A comparison of bismuth germanate, cesium iodide and sodium iodide scintillation detectors for gamma ray spectral logging in small diameter boreholes.
   J. C. Conway, P. G. Killeen, and W.G. Hyatt. Current research, part B, Geological Survey of Canada, paper 80-1B, p. 173-177 (1980)

   BGO offers a reduction of more than 50% in statistical errors in uranium determinations to NaI(Tl).

High Energy Physics

1. The response of a 5cm x 20cm BGO crystal to electrons in the 150 Mev to 700 Mev range.
   P. Pavlopoulos, M. Hasinoff, J. Repond et al submitted to Nuclear Instruments and Methods 197, 331-334 (1982).

2. Bismuth Germanate, a Novel Material for Electromagnetic Shower Detectors.
   G. Blanar, H. Dietl, E. Lorenz et al proc. EPS Intl. Conf. on High Energy Physics. Lisbon (1981).

The next Ball in HEP will be a BGO Ball.

Space Physics

1. A bismuth germanate - shielded mercuric iodide x-ray detector for space applications.
   J. Vallerga, G. R. Ricker, W. S. Schnepple, C. Ortale. IEEE. Trans. Nuclear Science NS-29, 151-154 (1982).

2. The use of an active coded aperture for improved directional measurement in high energy x-ray astronomy.
   A. Johansson, B. C. Beron, L. Campbell et al. IEEE Trans Nuclear Science NS-27, 275-380 (1980).

Low/Medium Energy Physics

1. Gamma ray response of a 38mm bismuth germanate scintillator
   A. E. Evans, Jr. IEEE Trans. Nuclear Science NS-27, 172-175 (1980).

2. Bismuth germanate scintillators as detectors for high energy gamma radiation.
   D. H. Drake, J. Nilsson, J. Faucett. Nucl. Instr. and Methods 188, 313-317 (1981).

This reference list is by no means complete and the omission of any pertinent paper is apologized.

Acknowledgment

The help provided by B. K. Utts and V. Berner in obtaining spectral data is greatly appreciated; as well as the generous donation of samples for study by J. Hietanen without which this investigation would not have been possible.

REFERENCES

1. R. Hofstadter: Phys. Rev. 74, 100 (1948).

2. R. Hostadter: IEEE Trans. Nucl. Sci. 22, 13-35 (1975).

3. R. L. Heath, R. Hofstadter and E. B. Hughes: Nucl. Instr. Methods, 162: 431-476 (1979).

4. M. T. Weber and R. R. Monchamp J. Appl. Phys. 44, 5496 (1973).

5. O. H. Nestor and C. Y. Huang: IEEE Trans. Nucl. Sci. NS-22, 68 (1975).

6. G. N. Hounsfield: Br. T. Radial, 46, 1916 (1973).

7. M. R. Farukhi: Proc. Workshop on Trans. and Emission CT. Korea 1978.

8. Z. H. Cho and M. R. Farukhi: J. Nucl. Med. 18, 840 (1977).

9. C. T. Thompson, Y. L. Yamamoto and E. Mayer: Positome II. (Abstract) J. Comput. Assist Tomogr. 2, No. 5 (1978).

10. M. R. Farukhi: Proc. Workshop Time-of-flight Assisted Tomography. Washington Univ. St. Louis 1982.

11. L. R. Carroll, G. O. Hendry and T. D. Currin: IEEE Trans. Nucl. Sci. NS-27, 1128-1136 (1980).

12. N. Nohara, E. Tanaka, T. Tomitami et al: IEEE Trans. Nucl. Sci. NS-27, 1128-1136 (1980).

13. M. R. Farukhi: IEEE Trans. Nucl. Sci. NS-29, 1237-1251 (1982).

14. H. Murayama, N. Nohara, E. Tanaka and T. Hayashi: Nucl. Instr. and Meth. 192, 501-511 (1982).

15. K. Takami, K. Ishimatsu, T. Hayashi et al: IEEE Trans. Nucl. Sci. NS-29, 534 - 538 (1982).

16. T. Yamashita, M. Ito and T. Hayashi: Proc. Intl. Workshop on Physics and Engineering in Medical Imaging, Pacific Grove, Calif. 1982.

17. K. Okajima, K. Takami, K. Ueda et al: Rev. Sci. Instr. 53, 1285 (1982).

18. R. Allemand, C. Gresset, J. Vacher: J. Nucl. Med. 21, 153 (1980).

19. N. A. Mullani, D. C. Ficke, M. M. Ter-Pogossian, IEEE Trans. Nucl. Sci. NS-22, 572 (1979).

20. M. Moszynski, C. Gresset, J. Vacher, R. Ordu Nucl. Instr. and Methods 188, 403 (1981).

21. H. Uchida, Y. Yamashita, T. Yamashita, T. Hayashi Proc. 1982 Nucl. Sci. Symp. Washington, D.C. 1982.

22. M. Moszynski, J. Vacher, R. Ordu. To be published in Nucl. Instr. and Methods.

23. A. E. Evans, Jr. IEEE Trans. Nucl. Sci. NS-27, 172-175 (1980).

24. D. M. Drake, L. R. Nilsson, J. Faucett. Nucl. Instr. and Methods 188, 313-317 (1981).

25. K. Takagi, T. Oi, T. Fukuzawa et al. J. Crystal Growth 52, 584-587 (1981).

26. H. Piltingsrud J. Nucl. Med. 20, 1279-1285 (1979).

27. S. E. Derenzo J. Nucl. Med. 21, 971-977 (1981).

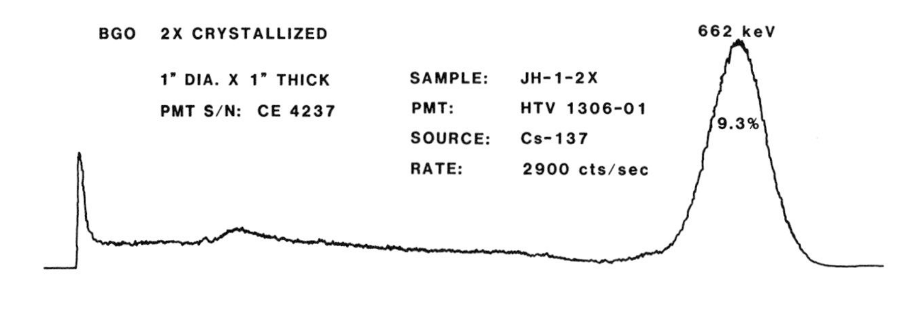

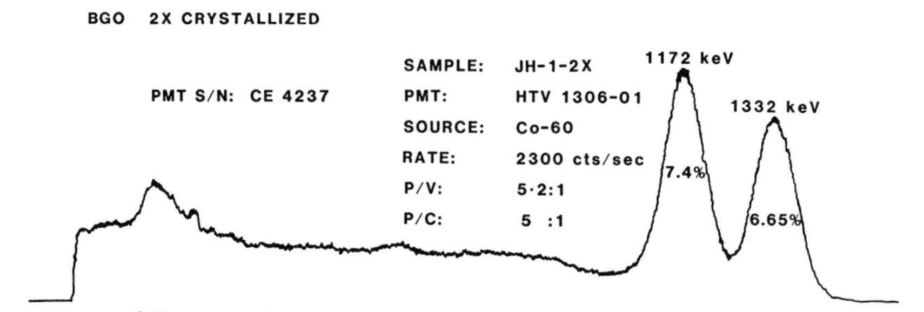

Figure 1. ^{137}Cs and ^{60}Co spectra for BGO (1" dia. x 1" thick). Sample JH-1-2X PMT 1306 S/N CE 4237.

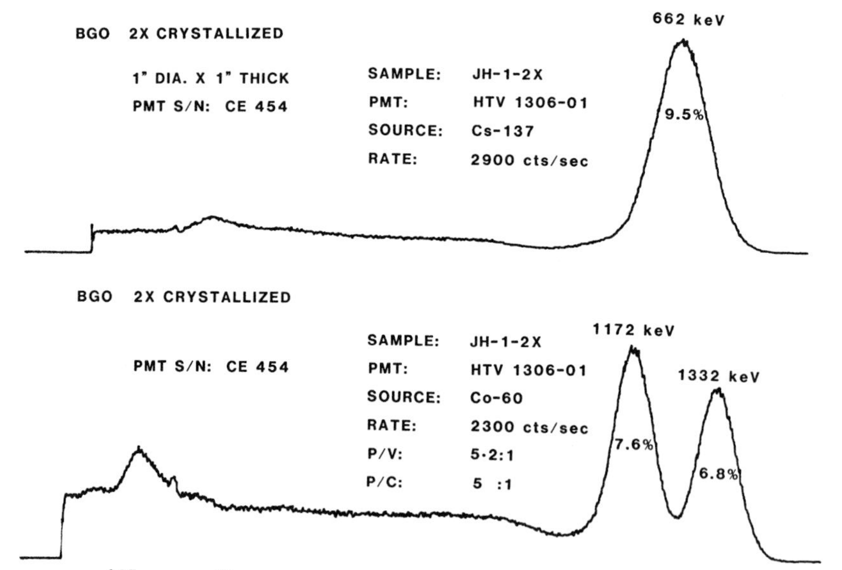

Figure 2. ^{137}Cs and ^{60}Co spectra for BGO repeated on S/N CE 454 same sample as Figure 1.

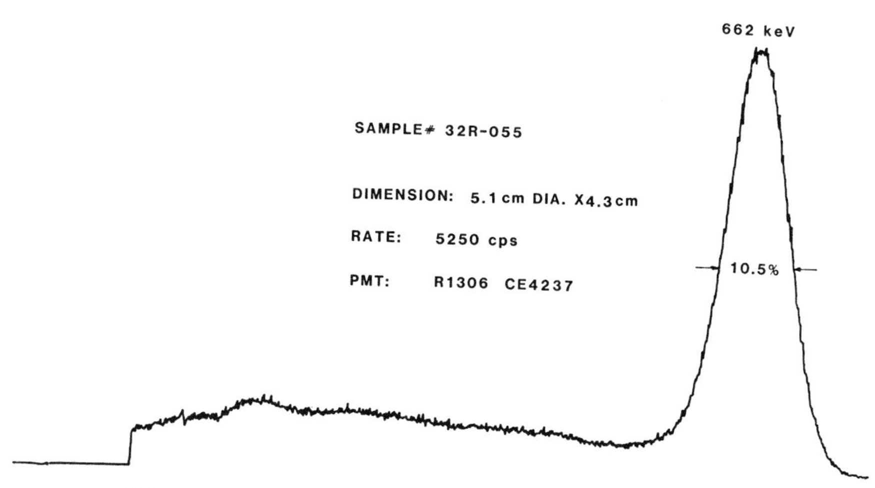

Figure 3. ^{137}Cs spectra for BGO (2.0X) sample 32 R-055. 51mm dia. x 43mm thick on HTV R1306 PMT.

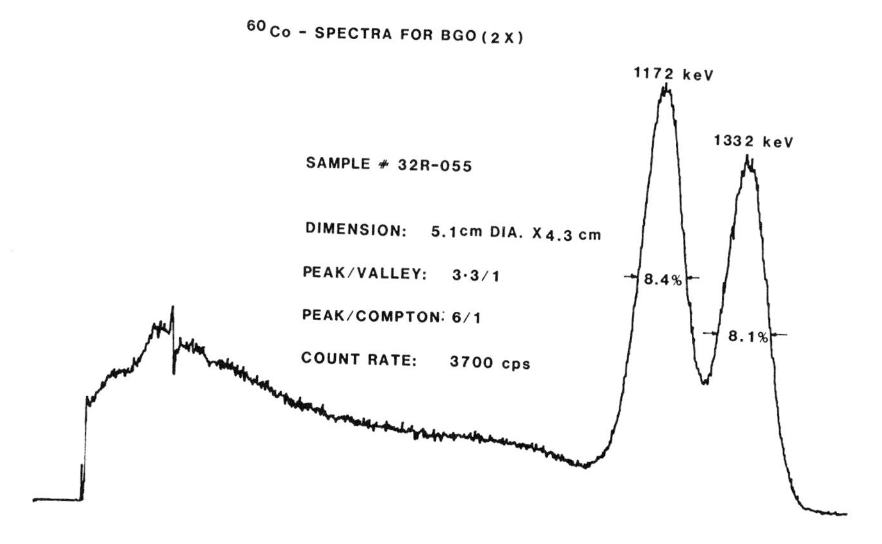

Figure 4. ^{60}Co spectra for 32R-055 BGO.

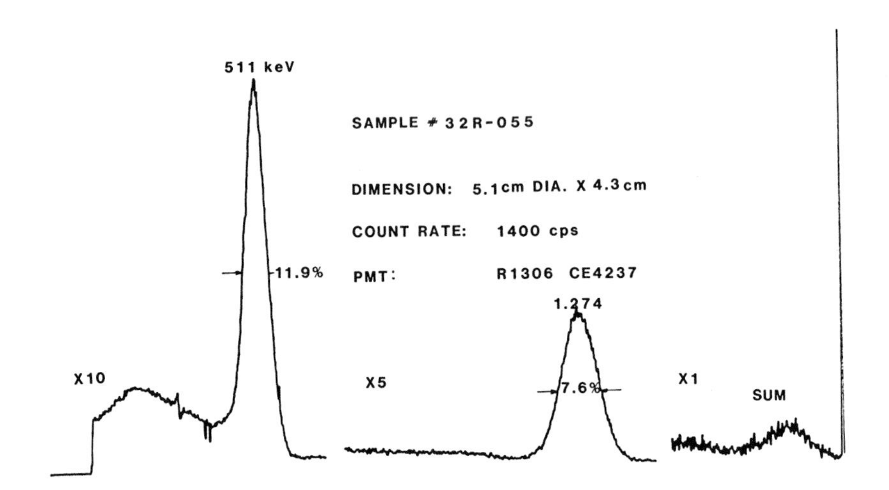

Figure 5. ^{22}Na spectra for 32R-055 BGO.

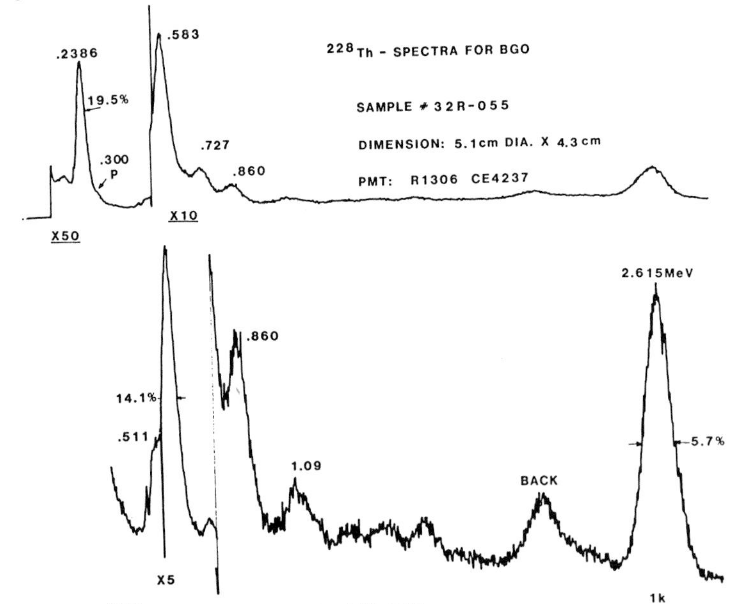

Figure 6. ^{228}Th spectra for 32R-055 BGO.

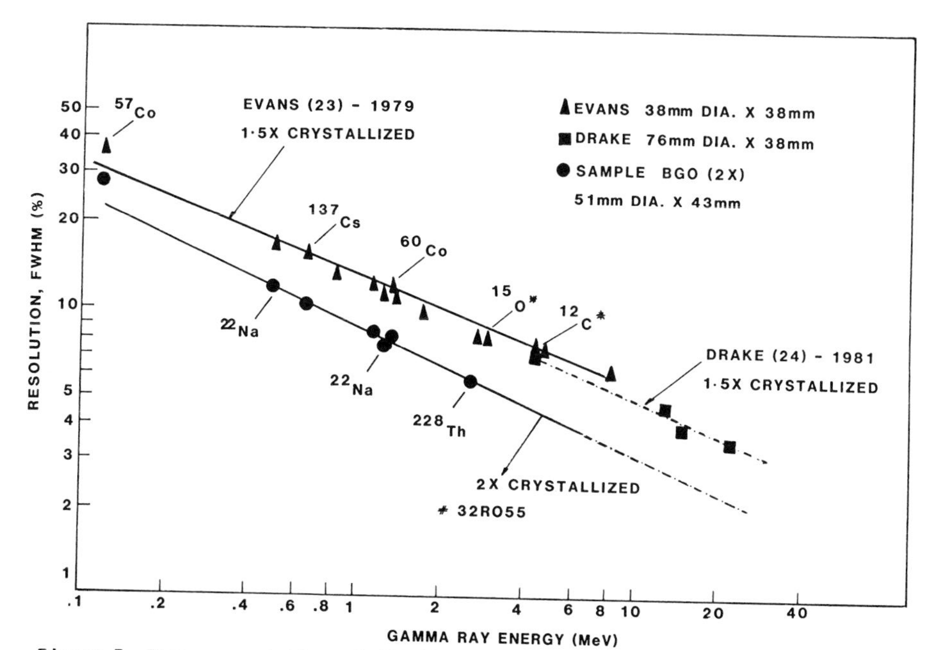

Figure 7. Energy resolution of 32R-055 BGO vs. gamma ray energy. Notice the overall improvement over (1.5X) grown material.

LARGE AREA SILICON AVALANCHE PHOTODIODES: PHOTOMULTIPLIER TUBE ALTERNATE

G. REIFF, M.R. SQUILLANTE, H.B. SERREZE AND G. ENTINE
Radiation Monitoring Devices, Inc., 44 Hunt St., Watertown, MA 02172 (617) 926-1167
GERALD C. HUTH
USC Medical Imaging Service, 4676 Admiralty Way, Marina del Rey, CA 90291

ABSTRACT

Silicon avalanche photodiodes have recently been shown to be a potential replacement for vacuum tube photomultipliers in many nuclear scintillation detector applications. The large active area, low noise, and ease of use of these solid-state photomultipliers makes them ideally suited to scintillation detector applications where overall size and ruggedness are a major concern. Historically, avalanche photodiodes have been limited for use in this capacity by small active areas, low internal gains, and poor optical sensitivity at the wavelengths at which most solid scintillator materials emit. Recent advances as the result of research aimed directly at the solution to these problems however, have successfully demonstrated one inch active area silicon avalanche photodiodes which produce a FWHM resolution of 9.5% for Cs137 at room temperature when coupled to a 1" x 1" NaI(Tl) scintillation crystal. Improvements to both material quality and device structure have advanced the state-of-the-art to make silicon avalanche photodiodes a viable alternative in scintillation gamma spectroscopy as well as for large area optical, beta, and low energy x-ray detectors.

INTRODUCTION

The traditional photomultiplier scintillator crystal combination has had great success over the years as a sensitive nuclear detector with moderate energy resolution. Its history has demonstrated that the performance of this detector system is quite good and it has become very valuable in a wide variety of applications. However, as the areas for scintillation spectroscopy increase and sophisticated technologies are demanding tighter and tighter restrictions on the size, weight, and versatility of detector systems, the inherent limitations of vacuum tube photomultipliers are becoming more obvious and, in many cases, burdensome. Specifically, some of the main concerns are exccessive bulk, inadequate ruggedness and a sensitivity to external magnetic and electric fields. These considerations play an important role in applications requiring shielded or collimated arrays such as in imaging systems, in satellite applications where size and weight can be a crucial factor, or in other areas where short warm-up time, transient voltage surges or rough handling significantly effects PM tube operation.

Mat. Res. Soc. Symp. Proc. Vol. 16 (1983) ©Elsevier Science Publishing Co., Inc.

The Si avalanche photodiodes recently developed have a large active area (one inch diameter) yet are extremely thin (0.4 mm) and lightweight. A scintillation detector utilizing such a device has the advantages of being very compact, highly sensitive, and is uneffected by external fields. In addition, being a true diode, the need for a bleeder string of resistors and capacitors is eliminated from the circuitry.

The solid-state nature of the avalanche photodiodes allows for application specific, custom designed fabrication. Although recent research has concentrated on a round geometry for coupling to standard NaI scintillator crystals, in theory both the size and shape may be varied to accomodate any particular counting geometry. For example, the medical imaging instrumentation used in positron emission tomography (PET) utilizes many scintillation detectors densely packed about a scanner. Positrons emitted by short lived isotopes within the patient's brain are detected and the signals intergrated to produce a computerized picture of the brain for analysis. For this application the scintillator most widely used is bismuth germinate (BGO) with a 2.0 cm x 0.5 cm window. Avalanche photodiodes have the potential to be custom fabricated to conform to the required shape of these crystals leaving minimal exposed active area. Coupling therefore is much more efficient, for unused active area contributes significantly to noise through excessive background counts and causing higher leakage currents. These devices would also have a significant advantage over PM tubes in this application with respect to their small size, since the density of the packing of the detectors is extremely important to resolution.

Enhanced short wavelength optical sensitivity with greater quantum efficiency than that of a typical photomultiplier tube, makes the development of high speed, multielement arrays for low light level detection a real possibility. Again, these could be extremely compact as many elements could be fabricated on a single silicon wafer.

THEORY OF OPERATION

The avalanche photodiode is very similar to other photodiodes but with some very unique characteristics.[1],[2],[3] Like other photodiodes, it is essentially a semiconductor material which contains a region depleted of charge carriers acting to electrically separate electron-hole pairs generated by incoming photons at the surface. The distinct difference lies in the ability of the avalanche photodiode to internally multiply the number of these electron-hole pairs, thus imparting a net gain on the incoming signal. The overall result is to transform an optical signal to an electrical signal, which is then amplified internally to a level where conventional electronics may then process this signal.

The diode is created by the formation of a slightly graded junction, with the diffusion of a dopant into a wafer of opposite semiconductor type. This gradient of carrier concentration from the wafer's surface to the junction imparts a slight electric field to this region of the device, which acts to direct the drift of charge carriers in the conduction band either toward or away from the p-n junction depending on their type. The operation of the device in the avalanching mode requires the application of a reverse bias (positive polarity to the n-type, or high donor level side) of over one thousand volts to instill a large electric potential across the junction. This bias should be large enough to cause impact ionizations, yet not so large as to cause high leakage current. Under these conditions the holes or electrons, (dependent on semiconductor type) generated at the surface by incident photons, drift under the influence of the dopant gradient to the space charge region where they quickly attain momentum large enough to cause collisions with bound electrons of the lattice due to the large electric field. These collisions are of the magnitude then to "knock" the hole-electron pairs into the conduction band thus freeing more carriers to be influenced by the electric field and cause further "knock-on" collisions with more bound electrons. The effect is to generate hundreds of free carriers from each charge carrier which drifts into the space charge region, thus achieving a net gain of electric signal. As shown diagramatically in figure 1, the active portion of the avalanche photodiode is then composed of two discrete regions; the drift region which is 10-60 microns thick, and the avalanche, or multiplication region which is about 200 microns thick. The total active thickness is around 300 microns (12 mils) so these devices may be relatively thin, yet thick enough to be mechanically strong in handling.

Depicted in figure 1 as well, is the edge bevel which is used to control surface fields which may cause premature breakdown due to surface currents. The bevel angle may range from 5-15° for optimal performance and is the most fragile aspect of the device since the thickness reduces to less than 2 mils at the front surface. However, given the proper encapsulant and packaging, this fragility can be overcome to make the device very rugged in use.

MATERIAL SELECTION AND DESIGN CONSIDERATIONS

The presently available,commercial silicon avalanche photodiodes, still possess severe limitations which make them inadequate for scintillation detection applications, as well as for some other optical and nuclear detector uses. Specifically, present SiAPDs[4] are limited to small active areas (less than 3 mm diameter), have low internal gains (about 60), and are insensitive to the short wavelengths where most popular scintillator crystals emit (410 nm for NaI(Tl) and 480 nm for BGO). These limitations become obvious in the following presentation of important design considerations. Efforts of the recent research to overcome these limitations are also described.

In scintillator detection applications, the necessary characteristics of avalanche photodiodes are well defined. They must be sensitive to the incoming wavelengths, have a large enough active area to allow coupling to crystals of sufficient size for high sensitivity, avalanche at a practical voltage level with sufficient gain for external electronics to easily resolve the signal, and they must have low currents and noise levels to achieve high signal-to-noise ratio necessary for good resolution.

Of prime importance in the design of avalanche detectors are those factors which may effect the noise of the device. The noise of avalanche detectors arises from three main sources as seen in figure 2. These consist of capacitance noise, shot noise, and noise arising from the statistics of the avalanching process itself. The capacitance noise consists of 1/f noise from the detector and preamplifier capacitances and series white noise. This noise may be optimized through preamplifier selection as well as selection of the base material and doping profile of the junction. Bulk resistivity of the base material also plays a role here, as this determines the voltage (V) necessary to achieve the breakdown condition. These considerations may be seen in the relation:

$$C \cong A\sqrt{q\epsilon N_D/2V} \qquad (1)$$

where A is area, ϵ, the dielectric constant, and N_D the carrier concentration.

The shot noise also consists of two components, internal current and surface current. The latter may be optimized by polishing the front surface to reduce leakage paths and by beveling the edges which spreads surface equipotentials across the exposed junction. The internal current, however, is dominated by the generation-recombination current within the space charge region and therefore is dependent on depletion layer width, as well as intrinsic carrier concentration and carrier lifetimes. Shot noise then, may be given as:

$$P_s = (qn_iW/\tau_e)\ A + P_s^1 \qquad (2)$$

Here n_i is the intrinsic carrier concentration, W the depletion layer width, and τ_e the effective carrier lifetime. P_s^1 is the surface current component of shot noise. As can be seen, shot noise is reduced by high carrier lifetimes within the space charge region, and as opposed to capacitance noise, increases with depleation layer width.

Thus, the importance of material selection and design begins to become evident, particularly with base resistivity. This one parameter not only defines the intrinsic carrier concentration which effects noise through the relations of equation (2), but it also determines the voltage necessary for the breakdown condition, which must be at a practical level for ease of use. In turn, the resistivity, doping profile, and voltage all effect the depletion layer width which defines the multiplication region of the device and determines device gain.

The third noise component, the excess noise, is perhaps the most important consideration in avalanche photodiode design. It may be thought of as the internal current multiplied by the effects of device gain. It has the relation:

$$P_E = P_s M^2 F \qquad (3)$$

where M is gain and F is the "excess noise factor" or:

$$F = kM + (1-k)(2-1/M) \qquad (4)$$

In the above equation k is the ratio of hole-to-electron ionization rates in the semiconductor. It is here where the selection of silicon becomes an important choice. Silicon inherently has a low k value;[5] electrons producing many more hole-electron pairs per unit distance through impact ionizations, when influenced under an electric field. As can be seen in equation (4) a low k value reduces the excess noise factor associated with the avalanche process, thus total device noise.

As well as reducing noise, the choice of Si for avalanche photodiodes for use in scintillation detection applications also takes advantage of its low stopping power. Because one is only interested in signals directly from the scintillator crystal, the low stopping power of the silicon will greatly reduce the observation of extra events occuring within the photodiode. This may even allow for illumination of the scintillator crystal through the photodiode when using gamma ray energies in excess of about 100 keV. Once Si has been selected, its properties then dictate the device geometry, that is, a p-type dopant diffused into a bulk n-type substrate. This configuration is superior to the inverse since electrons are then the primary carriers, which, for silicon have the higher ionization coefficient.

Although presently available silicon avalanche photodiodes take advantage of many of the above considerations, they still possess many limitations which required further design modifications to become successful large area scintillation detectors. One of the main shortcomings is an insensitivity to short wavelength light (See figure 3). This poor response towards the blue portion of the spectrum is due to an inactive or "dead" layer present at the surface of the device which may be as great as 1 micron deep. Photons absorbed within this region generate carriers which actually drift in the wrong direction, are lost to surface currents, and therefore not multiplied by the diode. The efforts of the recent research concentrated on reducing this inversion layer, resulting in a reduction to a tenth of a micron or less and therefore significantly enhancing short wavelength response.

The limitations on gain in these devices stem from the same problems which limit size. Although an increase in either of these two parameters necessarily increases noise, as seen by the noise relations presented earlier, this does not become a significant obstacle, up to a point. This can be demonstrated through an approximation:

$$M_O = (2/k + (P_c + P_s)/P_s)^{1/3} \quad (5)$$

where M_O is the optimal operational gain. Optimum gain then increases as the cube root of the noise, which increases with size and particularly, gain. Stated differently, M_O is the point at which noise begins increasing faster than the signal. The limitations on present silicon avalanche photodiodes is actually due to nonuniformity of the base material. Local avalanche breakdown is caused by regions of lower resistivity within the devices. This limits the gain to that level which is characteristic of the lowest electric field causing breakdown. By the same token, increasing size also increases the areas of nonuniformity improving the chances for premature breakdown and insufficient gain.

This problem has been overcome by using highly uniform transmutation doped silicon, made available relatively recently through improved technologies. The uniformity of this silicon is such that devices of many square centimeters are possible, achieving a large active area while experiencing the full gain potential of the device.

EXPERIMENTAL RESULTS

Silicon avalanche photodiodes fabricated in view of the material and design considerations as outlined in the previous section have been extremely successful in demonstrating large area solid-state scintillation detection. One inch diameter active area devices can typically have gains to over 300 when reversed biased near breakdown, as can be seen in Fig. 4. The optimum operational gain of the devices, as determined by equation (5) above, is approximately 130. This figure is further verified by examining Fig. 5. Here the relative noise voltage was measured using an RMS voltmeter and plotted as a function of device gain. At a device gain of 130 (corresponding to a reverse bias voltage of 1450V) the noise is seen to increase at a higher rate than the signal. This value of internal gain is more than sufficient to allow further amplification by the external electronics. This gain is possible with such a large area device due to the improved uniformity of the silicon used.

Although the gain of the avalanche photodiode does not approach that of a photomultiplier tube, the signal is sufficiently large that modern preamplifiers do not add to the noise. With enhanced short wavelength sensitivity, brought about by the reduction of the front surface dead layer, the signal is much higher prior to multiplication by the gain mechanisms. The gain therefore, in silicon avalanche photodiodes, need not be as high. Figure 3 compares the spectral response curve for the one inch SiAPD to both typical PM tubes and present commercially available, small area SiAPD's. As demonstrated, the short wavelength enhanced devices are much better matched to the output of NaI(Tl) scintillator crystals. Fig. 6 and 7 show Cs137 and Co^{57} spectra obtained with a 1" active area silicon avalanche photodiode coupled to a 1" NaI(Tl) crystal. With a device gain of 150 (optimum for this particular device), a linear amplifier gain setting of 100, and time constant of 0.5 usec., the FWHM resolution is seen to be 9.4% for Cs137 at room temperature.

In order to investigate the silicon avalanche photodiode as a large area, low energy radiation detector, the front surface was irradiated directly with an Fe55 x-ray source. The resulting spectrum is shown in figure 8. The detector was successful in resolving the Mn peak (5.9 keV) from the noise of the system.

The stability of these devices over time also appears very good with little or no deviation in performance over extended periods. In addition, initial temperature studies conducted with the devices demonstrated their ability to resolve Cs137 when coupled to NaI(Tl) to well above 50°C. Finally, the encapsulated and packaged large area silicon avalanche photodiode scintillation detector proves to be a very compact, rugged, sensitive detector which is extremely simple to use.

CONCLUSIONS

The results are very encouraging for scintillation detection utilizing large area silicon avalanche photodiodes. The advantages of these devices over conventional photomultiplier tubes have been shown to be numerous. Its small size combined with a large active area provide high sensitivity in a compact unit compatable with applications having stringent size-weight restrictions, such as those encountered in the satellite and medical imaging fields. In addition, the response uniformity and enhanced optical sensitivity, combined with internal gain, demonstrate the potential for multi-element, optical detector arrays for low light level detection. The devices are extremely versatile as exemplified by their ability to directly detect both high and low energy x-rays. Recent tests have also shown the detection of alpha particles from Am241 while being operated in the photovoltaic mode, that is, without external bias applied.

Although current devices may not match the resolution of present photomultiplier tubes, they are quickly approaching the point where the high energy performance with NaI is limited only by the scintillator crystal. The optical response can be further enhanced through the utilization of presently existing technologies which have emerged mainly as a result of various solar cell efforts. In so doing, the projections indicate that the signal can be further improved thus boosting the signal-to-noise ratio and increasing the resolution and sensitivity of the avalanche photodiodes.

REFERENCES

1. P. P. Webb, R. J. McIntyre, and J. Conradio, RCA Review 35, 234 (1974).

2. R. J. McIntyre, IEEE Trans. Electron Devices ED-19, 703 (1972).

3. "Research on Avalanche Type Semiconductor Radiaton Detectors", Gerald C. Huth, Peter V. Hewlsa and Vincent L. Gelezunas, Report Code NYO:3246TA-6 (Jan. 1970).

4. General Electric space Technology Products, (sales brochure data sheet) GE-STP (7-70).

5. S. M. Sze, Physics of Semiconductor Devices, Second Edition, Wiley-Interscience, new York (1981).

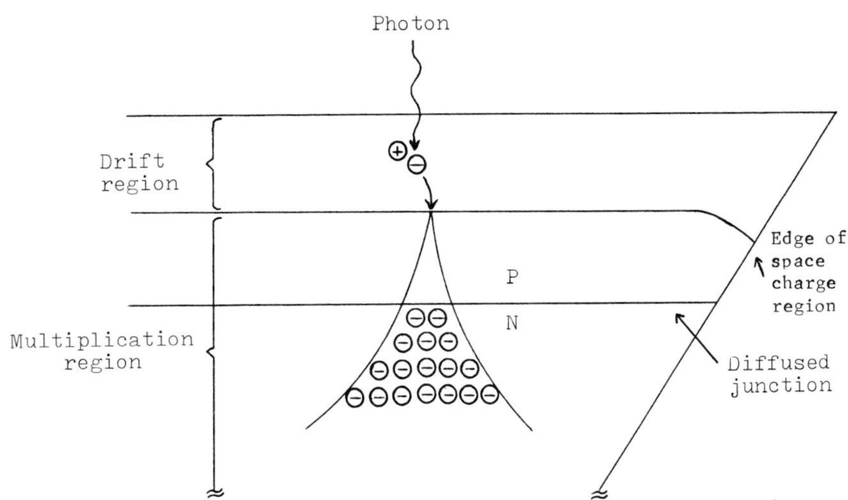

Fig. 1. Operation of internally amplifying detector.

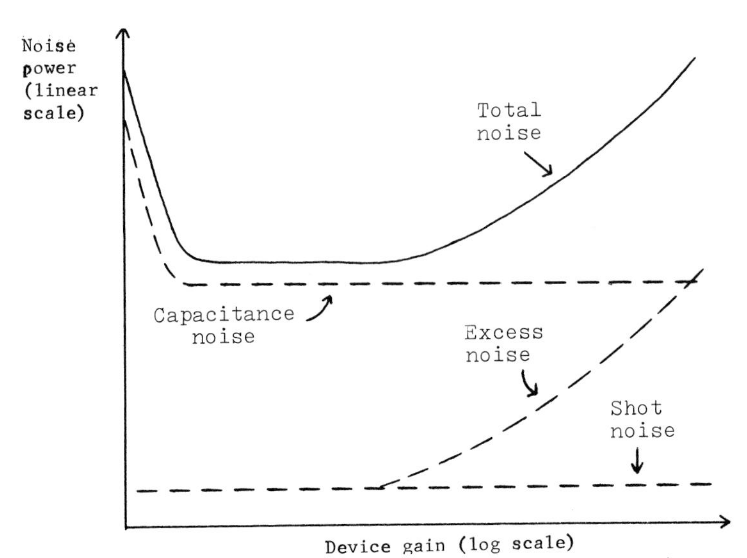

Fig. 2. Diagramatic illustration of noise effects in large area avalanche photodiodes.

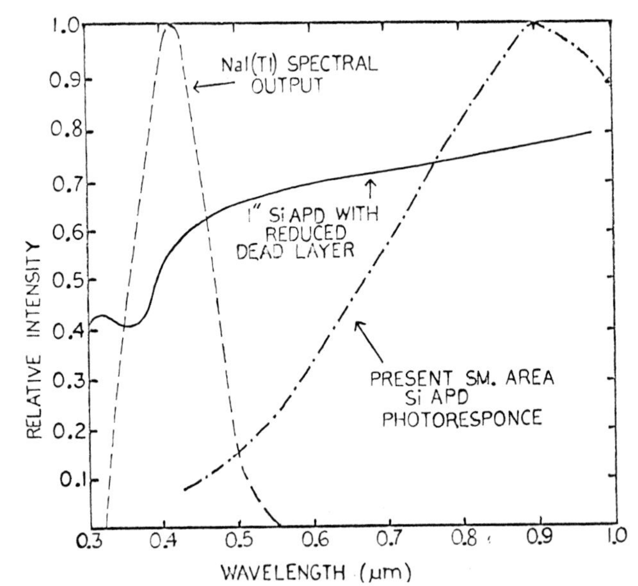

Fig. 3. Comparison of NaI spectral output to Si avalanche photodiode spectral response.

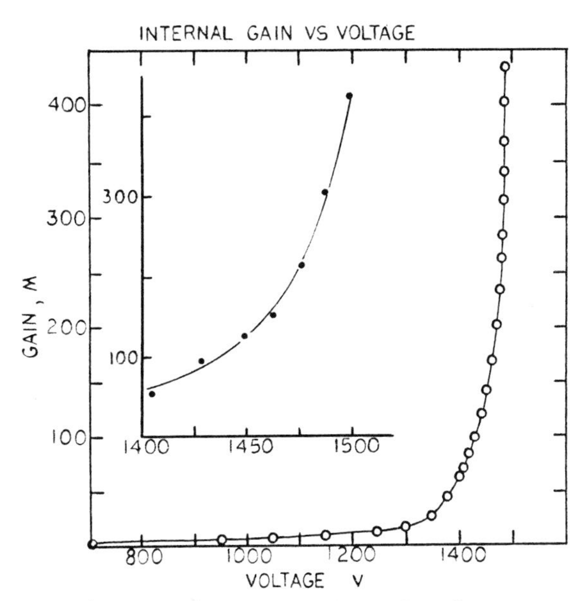

Fig. 4. Internal gain vs voltage for 1" Si avalanche photodiode.

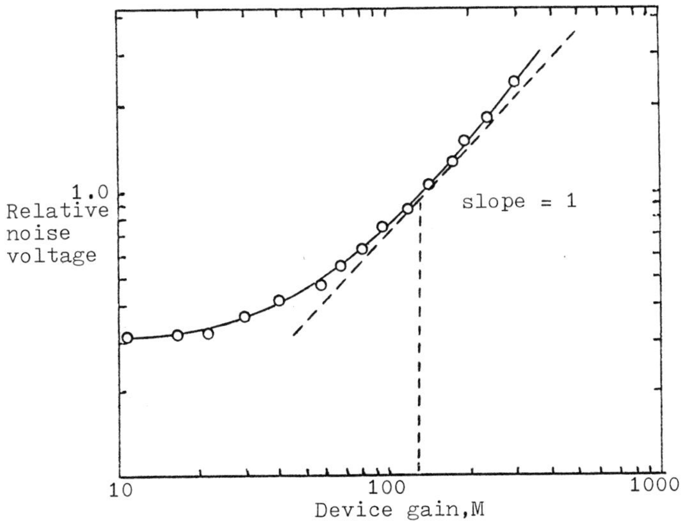

Fig. 5. Noise vs. internal gain for a one inch silicon avalanche photodiode.

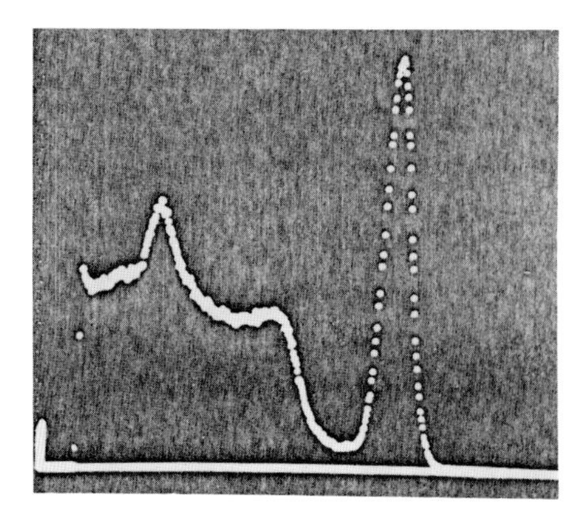

Fig. 6. Cs 137 pulse height spectrum for a 1" SiAPD coupled to a 1" NaI(Tl) scintillation crystal. FWHM is 9.4% at 25°C.

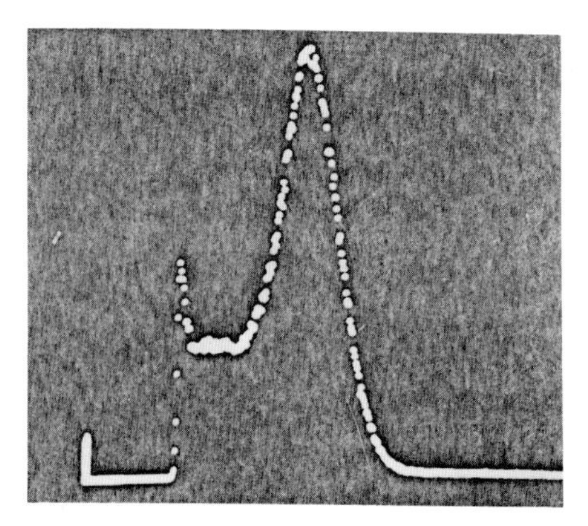

Fig. 7. Pulse height spectrum obtained from Co 57 irradiation of a 1" SiAPD coupled to a 1" NaI(Tl) scintillation crystal.

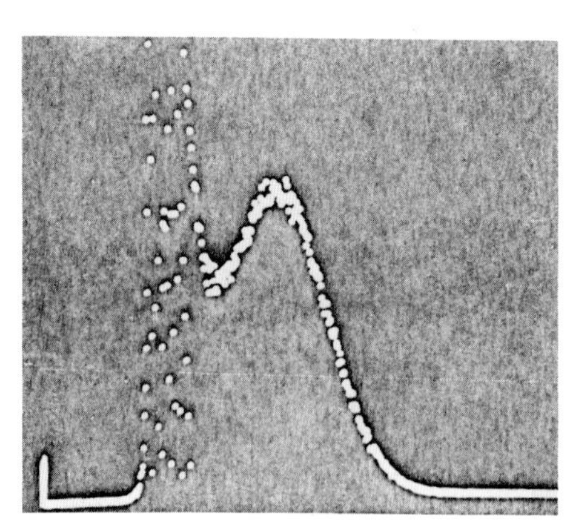

Fig. 8. Pulse height spectrum of large area SiAPD directly irradiated with Fe 55. (5.9 keV x-ray peak)

SILICON RADIATION DETECTORS - MATERIALS AND APPLICATIONS

JACK T. WALTON[1] AND EUGENE E. HALLER[1,2]
[1]Lawrence Berkeley Laboratory and [2]Department of Materials Science
University of California, Berkeley, CA 94720 U.S.A.

ABSTRACT

Silicon nuclear radiation detectors are available today in a large variety of sizes and types. This profusion has been made possible by the ever increasing quality and diameter silicon single crystals, new processing technologies and techniques, and innovative detector design. The salient characteristics of the four basic detector groups, diffused junction, ion implanted, surface barrier, and lithium drift are reviewed along with the silicon crystal requirements. Results of crystal imperfections detected by lithium ion compensation are presented. Processing technologies and techniques are described. Two recent novel position-sensitive detector designs are discussed--one in high-energy particle track reconstruction and the other in x-ray angiography. The unique experimental results obtained with these devices are presented.

INTRODUCTION

The principal semiconductor materials used in nuclear radiation detector fabrication are silicon and germanium. Germanium has benefited from an extensive program in its application to nuclear radiation detection. Silicon, however, has seen little work in this area. The reasons for this are readily cited:

- Nuclear structure studies which involve the detection of gamma rays have prompted the search for high atomic number detector materials and the improvement of germanium purification and crystallography has been a result of this "nuclear" interest. In addition, the lower processing temperature of germanium compared with silicon has permitted the purification and crystal growth of germanium for detector applications with a relatively modest investment.
- Silicon crystals with adequate material characteristics have been available from crystal suppliers to the semiconductor industry at large. However, the silicon processing is usually proprietary and consequently little dialogue has been possible between the detector maker and the crystal manufacturer.

However tenuous the supply of silicon for nuclear detector applications may appear, silicon is used extensively in the fabrication of nuclear radiation detectors. Unlike germanium detectors, silicon detectors can be operated at room temperature in many applications. Further, the variety of processing techniques which have been developed in part for integrated circuits fabrication, and the ever increasing size and quality of the silicon single crystals have led to a profusion of silicon detector types.

For the following discussion, let us divide silicon detectors into two classes. We base this division somewhat arbitrarily on the two kinds of silicon detectors which are typically used in charged particle identifiers (Figure 1) [1]. Thin detectors (5 - 500μm) in this application, are employed to measure the differential energy loss, ΔE, and are commonly called ΔE detectors. The thicker detectors (500 - 5000μm) are used to measure the total energy, E, and are called E detectors. Algorithms have been developed which relate the differential energy loss, ΔE, and the total energy, E, to the particle mass, M, and atomic number, Z. One algorithm for the generation of a particle identification (P.I.) signal is:

$$P.I. = (E + \Delta E)^n - E^n \propto T M^{n-1} Z^2 \quad (1)$$

where T is the ΔE detector thickness and n is a coefficient dependent somewhat on the particle (n = 1.73 for light ions).

The differences in the physical and electronic requirements on these two detector types, ΔE and E, are reflected directly in the fabrication processes. The ΔE detectors are normally characterized by:

- Thickness: 5 to 500μm
- Total volume active (totally depleted)
- Thin contacts on both sides
- Uniform thickness (± 1μm)

whereas the E detectors are characterized by:

- Thickness 500 - 5000μm
- Uniformity not critical
- Thin contact one side
- Partial volume active (partially depleted)

Depending on the specific application, there may be overlaps of the characteristic parameters between these two broad detector classes. Thick detectors have been used occasionally to measure the differential energy loss and, conversely, thin detectors have been used to determine the total particle energy. But for this discussion the concept of ΔE and E detectors will aid in illustrating why detector contacts, active volume, etc. are of interest.

In addition to their use at room temperature in charged particle detection, silicon detectors are also used extensively at cryogenic temperatures to detect low energy x-rays (1 - 30keV). The detectors employed here are normally thick, i.e., E detectors, since the efficient detection of the photons is of interest.

Finally, in addition to knowing the differential particle energy loss or the particle or photon energy, position information is frequently required. There exist many designs for both ΔE and E position-sensitive detectors. After considering the salient features of ΔE and E detector fabrication techniques in the following sections, two examples of position-sensitive detectors will be presented.

FABRICATION TECHNIQUES AND DETECTOR CHARACTERISTICS

Basic semiconductor detector diode structure

For the discussion to follow it will be useful to recall a few basic features of semiconductor diodes [2]. Figure 2 shows a schematic cross section of a planar n^+ i p^+ diode where the incident radiation, either photons (γ, x-rays) or charged particles (α, β, ions), produces tracks of electron-hole pairs. In the presence of the electric field as shown in the figure, the electron-hole pairs separate and rapidly drift to the contacts. The charge, Q_S, collected at the contacts is proportional to the energy, E, of the incident radiation:

$$Q_S = \frac{E}{\varepsilon} q \quad (2)$$

where ε_{Si} = 3.6eV, ε_{Ge} = 2.98eV and q = 1.6 x 10^{-19} coulombs. Further, the rms statistical fluctuation in Q_S, ΔQ_S is:

$$\Delta Q_S = \left(\frac{FE}{\varepsilon}\right)^{1/2} q \tag{3}$$

where F is the Fano factor which describes the deviation from normal statistics. The best value currently accepted for F ≃ 0.1 in both silicon and germanium. Values much greater than this are indicative of signal charge collection problems (i.e., trapping centers) in the detector material.

If a device as shown in Fig. 2 is to be an ideal radiation detector, the following set of requirements have to be met: a) the sensitive region must be free of charge trapping centers which would affect the amount of charge collected; b) the contacts need to be thin enough to allow the passage of radiation without appreciable energy or intensity loss; c) the contacts need to remain non-injecting for reverse bias field strengths of $E \simeq 10^2$ to $10^4 Vcm^{-1}$; d) the surfaces have to be passivated against detrimental effects from the ambients; and e) the noise which is caused by the reverse leakage current has to be kept small compared to the signal fluctuations. In the case of x-ray detectors, this leads to the requirement of cooling the detectors.

The signal charge, Q_S, produced by nuclear radiation and the rms fluctuation in this charge, ΔQ_S (Eqs. 2 and 3), as a function of energy are shown in Fig. 3. Also indicated in the figure are representative values of noise levels for a silicon detector and amplifying electronics achieved at room temperature and at cryogenic temperatures. The room temperature noise level is caused by shot noise resulting from a detector leakage current of about 5μA.

As shown in Fig. 2, there are two contacts on the detector, one of which is "rectifying" and the other which is often called a "blocking" contact*. In the case where the electric field lines from the "rectifying" contact reach through to the "blocking" contact (as in the case of the ΔE detectors) this "blocking" contact must inhibit the injection of minority carriers into the active volume of the detector. The formation of this "blocking" or "non-injecting" contact is as important for ΔE detectors' successful operation as the fabrication of the rectifying contact. With E detectors, the detector volume is frequently not totally active and consequently, the field lines from the rectifying junction may not reach the opposite contact. The requirement of full depletion of ΔE detectors and only partial depletion of E detectors and the resulting differences in contact properties and formation are the most distinguishing features between the two groups of detectors.

There are four basic technologies used to fabricate silicon ΔE and E detectors--three are related to contact formation, surface barrier (or metal Schottky barrier) diffused junction, and ion implantation, while the fourth lithium ion compensation, is concerned with the modification of the silicon material properties to allow wide depletion regions. Semiconductor detectors are frequently labeled by the dominant technology used in their fabrication even though more than one technology may be used. For example, the fabrication of lithium ion compensated diodes employs both diffusion and surface barrier technology. In the following sections the four basic detector types are separately reviewed. Special techniques employing epitaxial regrowth and multiple layered doped regions have also been explored, but they will not be discussed here.

---

*With an $n^+ \pi p^+$ structure, the n^+ is the rectifying contact and p^+ the blocking one. With an $n^+ \nu p^+$ the n^+ is blocking and the p^+ is rectifying. The symbols π and ν refer to high resistivity p and n material respectively. In the limit of ultra-pure or perfectly compensated material this distinction regarding the contacts vanishes.

Surface barrier detectors

A schematic of a silicon surface barrier detector and its holder is shown in Fig. 4. The following preparation steps are used to fabricate such a detector [3]. The silicon wafer after a series of chemical etching and surface preparation steps is potted in freshly mixed epoxy which has been painted onto a mounting ring*. The epoxies employed are either n-type (aliphatic amines) or p-type (catalytic) in their effects on the silicon surfaces. The details on how these epoxies affect the silicon surfaces and what influence the prior chemical treatment of the silicon wafer has on the resultant device performance are still not well understood. However, procedures have evolved which do allow the successful production of these detectors with the cured epoxy providing both the required junction protection and the mechanical adhesion to a mounting ring.

After curing the epoxy at 50 - 60°C, metal contacts (100 - 200Å thick) are evaporated over the epoxy and silicon surfaces. On the device shown in the figure, the aluminum forms the "rectifying" contact on the p-type silicon, while gold forms the noninjecting or "blocking" contact. With n-type surface barriers, the roles of these metals are reversed.

The contacts which are formed by evaporation of metals onto silicon surfaces which have been exposed to various oxidizing environments, do not adhere very strongly to the silicon. Consequently, the metal Schottky barrier contacts are easily damaged. Such damage can strongly affect both the rectifying and blocking properties of the contacts.

The silicon used in the fabrication of these detectors is extremely pure in comparison with that employed in the semiconductor electronics industry. For silicon the depletion region width, x_0, is given by [4]:

$$x_0(\mu m) = 3.59 \times 10^{-3}\left(\frac{V}{N}\right)^{1/2} \tag{4}$$

where N is the net impurity concentration (cm^{-3}) and V is the applied voltage (V). Figure 5 shows the depletion region width with 100 volts bias as a function of the net impurity concentration. From this figure, it is seen that the net impurity range of interest for ΔE detector fabrication is $[N] \leq 10^{15} cm^{-3}$ and for E detectors is $[N] \leq 3 \times 10^{11} cm^{-3}$.

The leakage current associated with a reverse biased semiconductor junction has three components: diffusion (or thermionic emission in the case of surface barriers), thermal generation and surface leakage. For silicon detectors operated at room temperature (300K) and lower, the diffusion component is so small that it can be ignored. The surface leakage current is strongly dependent on the fabrication technique, while the thermal generation current, i_g, at room temperature is given by [4]:

$$i_g = \frac{0.128\ x_0}{\tau}\ \mu A cm^{-2} \tag{5}$$

where the depletion depth, x_0, is measured in μm and τ is the minority carrier lifetime in μsec. From this it becomes clear that the numerical value of the minority carrier lifetime should be comparable to the numerical value of the detector thickness if the generation current is to be maintained at the μA level. High quality silicon crystals typically exhibit lifetimes of $\tau > 500 \mu sec$. Since the surface barrier detector fabrication does not involve

*While the figure shows a p-type silicon wafer, the more common commercially available surface barriers are made on n-type silicon.

any high temperature processing steps which could degrade the lifetime, the lifetime in the final detector should be that of the starting crystal and consequently the generation current should be less than 1 μAcm^{-2}.

The surface barrier detector material requirements and device characteristics are tabulated in Table I. Surface barrier detectors are fairly readily fabricated and have exceptionally thin contacts. They are, however, relatively fragile and are normally limited to readily machinable geometries.

Diffused junction detectors

Silicon planar diffused junction technology employing either n- or p-type silicon has been utilized for many years in the fabrication of silicon radiation detectors [5]. Figure 6, which shows a detector made on p-type silicon, indicates the features of these devices. The silicon dioxide grown at 950 - 1000°C and 3000 - 7000Å thick provides the junction diffusion mask and edge protection. Unlike the epoxy-silicon interface of the surface barrier detector, the properites of this silicon-silicon dioxide interface are fairly well understood [6]. The n^+ and p^+ contacts are diffused 0.2 - 0.5µm deep (typically at 900 - 950°C, 10 - 30 minutes) into the wafer and consequently are much more rugged than the metal contacts on the surface barrier detectors. The diffused contacts are, however, thicker than the metal barriers, and for many applications this is a serious limitation. In addition, the high temperature processing required for the growth of the silicon dioxide often causes a significant decrease in the minority carrier lifetime. This results in the generation current (Eq. 5), which is inversely proportional to the lifetime, being larger than that obtained with the surface barrier technology. However, the recently reported use of chlorine during the silicon dioxide growth has indicated that the minority carrier lifetime can be maintained and consequently the leakage currents reduced to values approaching that achieved with surface barrier detectors [7].

The processing of diffused junction detectors is complicated both in the procedures and equipment required in comparison with the surface barrier technology. However, diffused junction detectors are very rugged and the planar technology allows the ready use of photolithographic techniques which permit a wide variety of detector geometries to be produced. The diffused junction material requirements, commonly employed diffusants and detector characteristics are given in Table I.

Ion implanted detectors

Ion implantation is also used to fabricate the contacts on silicon radiation detectors. This technique offers the advantages of precise control of the depth distribution of the doping impurities and lower temperature processing in forming the contacts. Figure 7 indicates the features of a detector made on p-type silicon with ion implanted contacts. The silicon dioxide, as shown, again forms a mask and passivates the surface around the rectifying junction.

Processing normally involves the growth of silicon dioxide, as previously, followed by the implantation of the n^+ and p^+ contacts (10 - 30keV, 2 - 5 x 10^{12} ions cm^2). The radiation damage produced by the implantation is then annealed (~ 700 - 800°C, 10 - 30 minutes). Ion implanted contacts exhibit thicknesses of the order of 300 to 1000Å.

The ion implantation process does not require the same care which is needed for the diffusion process since the activation of dopants occurs at lower temperatures than those needed for diffusion. The doping impurities are introduced at room temperature or lower with very precise control. The equipment requirements are, however, greater. The use of ion implantation has resulted in devices which are rugged yet possess thin entrance contacts. The common dopants, material requirements and characteristics of these devices are also presented in Table I.

Lithium ion compensated detectors

While silicon crystals are available with resistivities which would allow the fabrication by the preceding techniques of detectors greater than 1mm in depletion depth, these crystals are presently expensive and scarce. Consequently, lithium ion compensation is extensively used to produce the thicker detectors. The following procedure is used for their fabrication. Lithium is diffused into the p-type wafer (nominal 1000 ohm-cm) forming the n^+ contact. A groove is then cut into the wafer using an ultrasonic cutter. This groove and the p^+ contact region are chemically etched with the resultant device as shown in Fig. 8. The p^+ contact is formed by evaporating gold onto the etched region. After surface treatment to minimize the leakage currents, the device is put onto "drift". The "drift" consists of applying a reverse bias (500 - 1000 volts) and heat (110 - 150°C) to the device. This causes the lithium ions to drift under the influence of the applied electric field from the n^+ contact into the bulk p material. The lithium ions, which are interstitial, compensate the nascent boron impurities to produce a nearly intrinsic material in the compensated region.

Lithium ion compensation has been used for many years in both silicon and germanium detector fabrication with varying degrees of success. A major obstacle to successful lithium ion compensation has, historically, been the presence of oxygen in the silicon and germanium crystals . Oxygen forms an immobile donor complex with lithium which leads, for medium oxygen concentrations ($[O_2] \simeq 10^{15}cm^{-3}$), to a reduction of the lithium ion mobility and in the case of high oxygen concentration ($[O_2] \geq 10^{17}cm^{-3}$), to a complete immobilization of the lithium for all practical purposes [8].

Another interference in the drift process has been studied more recently. It is related to microdefects (A- and B-type swirls) which can be present in large diameter, dislocation-free floating-zone crystals. The generation and distribution of the A and B swirls has been associated with the crystal growth conditions and crystal growing procedures have been advanced to reduce or eliminate their generation. Further, it is believed that A swirls, which consist of dislocation loops surrounding interstitial silicon atom platelets, evolve from B swirls which are partial loops [9].

We have on occasion encountered dislocation-free silicon crystals in which the lithium ion mobility was reduced in spite of the fact that the oxygen concentration was low (i.e., $[O_2] < 10^{15}cm^{-3}$). Consequently, we initiated a study to determine whether the A and B swirls which are affected directly by the growth conditions of dislocation-free silicon were responsible for the observed lithium ion mobility and/or the charge collection in the completed detectors.

Our study [10] consisted of five 1000 - 2000 Ωcm p-type crystals which had been grown under different floating zone conditions. Two crystals were grown with a relatively large molten zone, while the remaining three were grown with a smaller zone (the large zone was 50% greater than the small one). The lithium compensated detectors made from these crystals were scanned with an alpha source with various applied bias voltages. The results of these alpha scans are shown in Fig. 9. It is readily apparent that there is a marked difference in the low bias charge collection properties of these detectors. The detectors made from the two crystals with the larger molten zone have constant charge collection response, whereas those made from the crystals with the smaller zone show pronounced dips in the central region. We believe that this charge collection deficiency as shown in Fig. 9 is related to the formation of micro defects (pre B swirls) during the crystal growth. Further, we believe that if these defects are present in sufficient density, the lithium ion mobility will be reduced. This technique of examining lithium compensated detector charge collection performance with low bias voltages appears to be very sensitive to the presence of micro defects in the crystal. While this has proven useful in assessing the quality of silicon for detector fabrication, it may also be useful in correlating crystal growth conditions with defect formation.

The silicon crystals employed in the lithium ion compensation technique are of the floating zone kind, p type with 1000 - 2000Ωcm resistivity (corresponding to $[B] \simeq 10^{13}cm^{-3}$) and 500 - 1000μsec minority carrier lifetime. The resistivity range is limited at high boron concentrations ($[B] \geq 10^{14}cm^{-3}$) by the reduced lithium ion mobility due to lithium boron pair formation. Low boron concentrations ($[B] \leq 5 \times 10^{12}cm^{-3}$) make it difficult to control the silicon surface states and, consequently, the leakage currents during the ion drift process.

The presence of dislocations also affects the lithium ion mobility [11]. Consequently, we normally use crystals having a dislocation density between 0 and $4000cm^{-2}$. However, we noted in our previously mentioned work on dislocation-free silicon crystals that the presence of "pre-B" type swirls in sufficient density can also reduce the lithium ion mobility. Therefore, the specification of zero dislocations may not necessarily guarantee an adequate lithium ion mobility in the crystal.

The specification of a minority carrier lifetime of 500 - 1000μsec is intended to select crystals which have a small deep trap concentration. Deep traps affect the extremely stringent charge collection properties of nuclear radiation detectors. Mayer [12] has shown that for the trapping effects to be negligible , the mean time that a carrier must be free before being trapped, τ^+, is related to the detector thickness, L, by:

$$\tau^+ = \frac{100L}{v_d} \qquad (6)$$

where v_d is the carrier drift velocity, ~ $1 \times 10^6 cms^{-1}$, in the usual lithium drift detector at room temperature. For a 5mm thick detector, $\tau^+ = 50\mu sec$. The parameter τ^+ is a detector parameter and is not the minority carrier lifetime which is measured at near equilibrium conditions. The selection of 500 - 1000μsec for the minority carrier lifetime specification avoids any charge collection problems which could be due to this crystal parameter.

While the preceding discussion has centered on room temperature applications of these detectors, lithium drifted detectors are also used extensively at cryogenic temperatures to detect soft x-rays (E = 1 - 30keV). In these applications, trapping centers which are inactive at room temperature become much more critical. However at the present time, there is little understanding of the nature and density of the traps and their influence on the charge collection processes at these low temperatures. Fano factors for lithium compensated silicon detectors of the order of 0.1, which is indicative of low trap density (see discussion regarding Eq. 3), are commonly obtained so that the anomalous crystals have not yet warranted investigation.

The junction edge protection on lithium drifted detectors is frequently accomplished by coating the surfaces with paints, varnishes or other materials which do not substantially alter the detector surface states. The exact formulation of these coating materials employed by the various detector manufacturers is proprietary as it has normally been developed over a long series of trials and evaluations.

The lithium compensation can be done with little equipment and the process itself is comparatively simple. Passivation of the device surfaces, however, borders on being an art and consequently the long term stability of these devices, especially at room temperature, is variable. Nevertheless these detectors are used extensively in ΔE, E telescopes and in x-ray detector systems. The main features of these devices are also given in Table I.

SILICON DETECTOR APPLICATIONS

In choosing the folowing examples to illustrate silicon detector capabilities, we attempted to select a few applications which not only indicate what has been accomplished, but which could also point out areas of future development.

ΔE, E telescopes

Detector telescopes employing ΔE, E dual arrangement as shown earlier in Fig. 1 or having multiple (three or more) detectors have been used for many years in the identification of nuclear particles. In the ΔE, E dual telescope a single identification is made using an algorithm such as that of Eq. 1. By employing a second ΔE detector and performing two identifications on the same particle passing through the telescope, the signals can be compared and anomalous events rejected. The performance of a dual telescope in measuring the products from 129MeV alpha particles impinging on a ^{12}C target are shown in Fig. 10a, while the improvement obtained by employing a ΔE,ΔE,E triple telescope during the same experiment is evident in Fig. 10b [13]. The particle identifier techniques have been extended to multiple-element detector telescopes which have been used to study cosmic ray composition [14] and neutron rich light nuclei stability [15]. In these latter applications, lithium ion compensated detectors with areas of 15 to 40cm^2, 3 to 5mm thick and 8 to 16 detectors in a telescope have been employed to measure the high energy, low intensity particles. Detectors of this area and thickness place very heavy demands on the silicon crystal quality and the future availability of such large area detector telescopes is very dependent on the interest and ability of crystal suppliers to provide these large diameter, high-purity silicon crystals. We have indicated previously the difficulties in lithium ion compensation with these large diameter crystals and additional work is required in relating the crystal growth conditions for these large crystals with the resultant detector characteristics.

Position-sensitive detectors

It was noted earlier that in addition to the particle energy, its position is also frequently of interest. Two basic types of semiconductor position-sensitive detectors are in use--continuous element and discrete element [16]. Continuous-element detectors use a resistive layer to generate the position signal. Discrete-element detectors have a series of what may be considered individual detectors on a common wafer. The following two examples are discrete-element devices, the first being used in high-energy physics and the second in nuclear medicine.

Particle track reconstruction: With the new high-energy physics accelerators proposed or under construction, there is a growing awareness that new generation of particle detectors is required to adequately utilize the capabilities of these machines. Among the new detectors proposed are arrays of silicon wafers 100 - 300μm thick, each having 50 - 500 equally spaced stripes (10 - 50μm wide, 10 - 100μm apart). In some applications, a small number of these detectors would be placed near the interaction region and would be used to project the particle tracks back into the interaction region or vertex where short-lived, charmed-particle decay (which is of interest) occurs.

A prototype of one of these vertex detectors is shown in Fig. 11. The detector is a 150μm thick diffused junction device having 48, 1cm long stripes, 20μm wide and 60μm apart.

The preliminary results obtained in a test at the Fermi Laboratory indicated that a five-detector telescope of these detectors was successful in particle track reconstruction [17]. Similar results have been obtained earlier at CERN [18] and we expect that detectors of this type will be in growing demand especially with the newer accelerators. There exists, however, a readout problem. The detectors tested at the Fermi Laboratory have one amplifier chain connected to each stripe. For experiments involving larger area devices and/or devices with a higher density of stripes, the electronics costs of this approach may become prohibitive. Consequently, development of these vertex detectors integrated with the required electronics or a ready hybridization arrangement appears to be of paramount importance.

X-ray angiography: Angiography [19], as currently employed, involves the visualization of blood vessels by use of the x-ray absorption characteristics of iodine-containing compounds which are injected directly into the artery of interest. Conventional x-ray sources provide a broad spectrum of energies and consequently, a relatively high dosage of iodine is required to obtain adequate images. For many individuals, the iodine concentration required and its injection by means of a catheter present serious risks. However, the availability of high-intensity x-rays produced by synchrotron radiation at SLAC allows the monochromatization of these x-rays with sufficient intensity that images can be produced with much lower concentrations of iodine and non-catheterization. The essence of the technique is shown in Fig. 12 where the monochromatization of synchrotron radiation by Bragg diffraction is indicated. The x-ray energy is switched above and below the K absorption edge of iodine to allow the subtraction of the background. While a 256-element detector is shown in the figure, preliminary measurements have, in fact, been performed with a 30-element detector shown schematically in Fig. 13. Results obtained on the blood flow in a dog's heart are shown in Fig. 14. Development of larger area detectors with smaller element size is now under way with the prospect of achieving high-quality images on heart patients as the immediate goal.

SUMMARY

Silicon radiation detectors have been examined principally in relation to their use in charged-particles telescopes. The four basic fabrication techniques, surface barrier, diffused junction, ion implantation and lithium ion compensation have been reviewed.

While silicon detectors are widely used in scientific research, the silicon crystals used to fabricate these detectors are produced by companies whose principal interest is supplying silicon to the semiconductor industry. Consequently, the silicon supply for detector fabrication is often tenuous. However, given the availability of high quality silicon the examples given demonstrate the flexibility of silicon detectors both in fabrication techniques and in final applications. Furthermore, the continued interest by the semiconductor industry in silicon processing will undoubtedly yield additional processing advances which will present even greater flexibility in silicon detector designs and uses.

ACKNOWLEDGMENTS

We appreciate the continued interest and comments of F. S. Goulding in regard to silicon semiconductor research and development. This review has also benefited from discussions with W. L. Hansen, R. H. Pehl, A. C. Thompson, H. A. Sommer and Y. Wong at LBL, E. B. Hughes at Stanford University and G. R. Kalbfleisch at the University of Oklahoma.

This work was supported by the Director's Office of Energy Research, Division of Nuclear Physics, and by Nuclear Sciences of the Basic Energy Program, Office of Health and Environmental Research of the U. S. Department of Energy under Contract No. DE-AC03-76SF00098.

REFERENCES

1. F. S. Goulding and B. G. Harvey, Annual Review of Nuclear Science 25, 167 (1975).

2. E. E. Haller, IEEE Trans. Nucl. Sci. NS-29, No. 3, 1109 (1982).

3. R. C. Trammell, IEEE Trans. Nucl. Sci. NS-25, No. 2, 910 (1978).

4. A. Coche and P. Siffert in: Semiconductor Detectors, G. Bertolini and A. Coche eds. (North-Holland, Amsterdam 1968) Ch. 2.

5. F. S. Goulding, Nucl. Instr. and Meth. 43, 1 (1966).

6. B. E. Deal in: Semiconductor Silicon 1977, H. R. Huff and E. Sirtl eds. (The Electrochemical Society, Princeton 1977) p. 276.

7. J. Kemmer, Nucl. Instr. and Meth. 169, 499 (1980).

8. P. Siffert and A. Coche in: Semiconductor Detectors, G. Bertolini and A. Coche eds. (North-Holland, Amsterdam 1968) Ch. 1.

9. A. J. R. DeKock in: Handbook on Semiconductors, Vol. 3, S. P. Keller ed. (North-Holland, Amsterdam 1980) Ch. 4.

10. A. Fong, J. T. Walton, E. E. Haller, H. A. Sommer and J. Guldberg, Nucl. Instr. and Meth. 199, 623 (1982).

11. H. J. Guislain, W. K. Schoenmaeker and L. H. DeLaet, Nucl. Instr. and Meth. 101, 1 (1972).

12. J. W. Mayer in: Semiconductor Detectors, G. Bertolini and A. Coche eds. (North-Holland, Amsterdam 1968) Ch. 5.

13. J. Cerny, S. W. Casper, G. W. Bulter, H. Brunnader, R. L. McGrath and F. S. Goulding, Nucl. Instr. and Meth. 45, 337 (1966).

14. M. E. Wiedenbeck and D. E. Greiner, The Astrophysical Journal 239, L139 (1980).

15. T. J. M. Symons in: Atomic Masses and Fundamental Constants, J. Nolen and W. Benenson eds. (Plenum Press, 1980) p. 61.

16. E. Laegsgaard, Nucl. Instr. and Meth. 162, 93 (1979).

17. G. R. Kalbfleisch, to be published in Nucl. Instr. and Meth.

18. E. H. M. Heijne, L. Hubbeling, B. D. Hyams, P. Jarron, P. Lazeyras, F. Pinz, J. C. Vermeulen and A. Wylie, Nucl. Instr. and Meth. 178, 331 (1980).

19. A. C. Thompson, F. S. Goulding, H. A. Sommer, J. T. Walton, E. B. Hughes, J. Rolfe and H. D. Zeman, IEEE Trans. Nucl. Sci. NS-29, No. 1, 793 (1982).

TABLE I.

Detector Type Parameter	Surface Barrier	Diffused	Ion Implanted	Lithium Ion Compensated
Resistivity (Ω cm)	50-50,000	50-50,000	50-50,000	1000-3000
Lifetime (μsec)	>100	>100	>100	>500
Type - Floating Zone	n, p	n, p	n, p	p
EPD (cm^{-2})	<4000	<4000	<4000	<4000
n^+ Contact	Al	P, As	P, As	Li
p^+ Contact	Au, Pt Pd, Cr, Ni	B	B	Au, Pt Pd, Cr, Ni
Contact Thickness ($\mu gmcm^{-2}$)	~20-40	~100-150	~60-80	p^+~20-40 n^+~200→ $3x10^4$
Thickness Range (μm)	5-1000	5-1000	5-1000	500-5000
Area (cm^2)	<30	<20	<30	<40

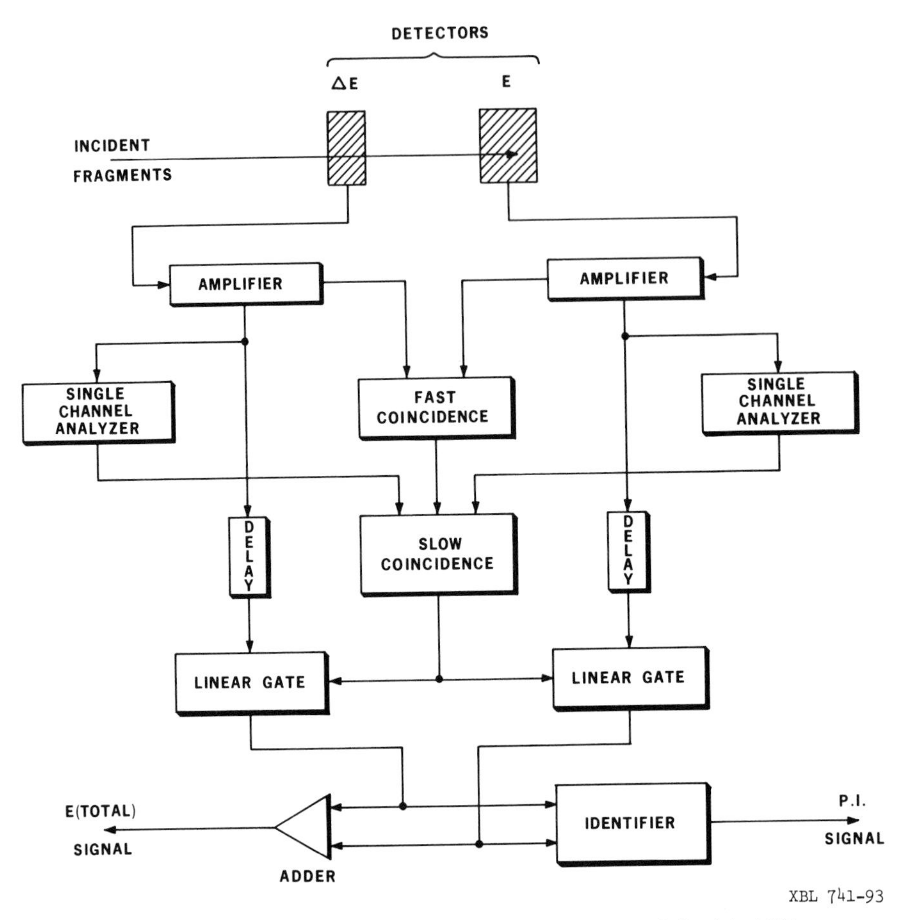

Fig. 1. Block diagram of the ΔE, E detector telescope particle identifier system typically employed to study nuclear reaction products.

+
n^+
$\vec{E}$
ν OR π
e_{ph}
p^+
γ,X
-
α,e$^-$

XBL 8210-2743

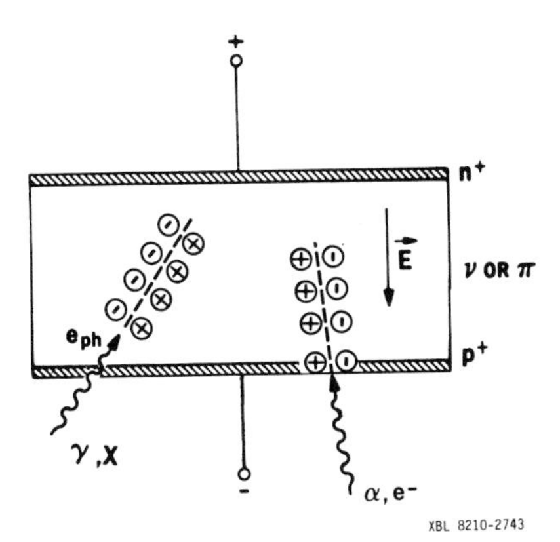

Fig. 2. Schematic cross secion of a nuclear radiation detector. The contacts are labeled n^+ and p^+ while the bulk material is labeled ν and π corresponding to high resistivity n or p silicon respectively.

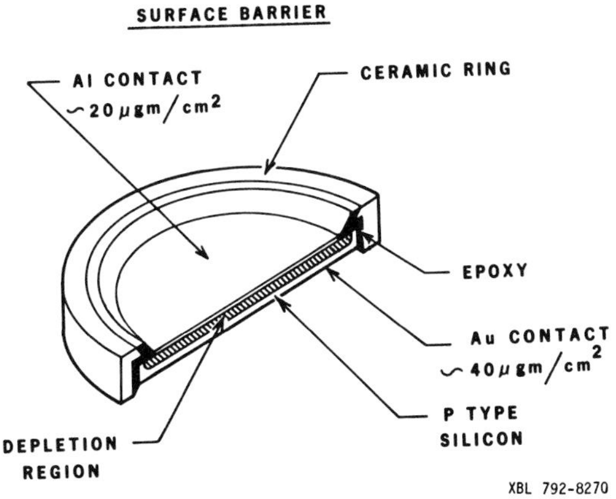

SURFACE BARRIER
Al CONTACT
~20 μgm/cm²
CERAMIC RING
EPOXY
Au CONTACT
~40 μgm/cm²
P TYPE
SILICON
DEPLETION
REGION

XBL 792-8270

Fig. 4. Mechanical details of surface barrier detector fabrication.

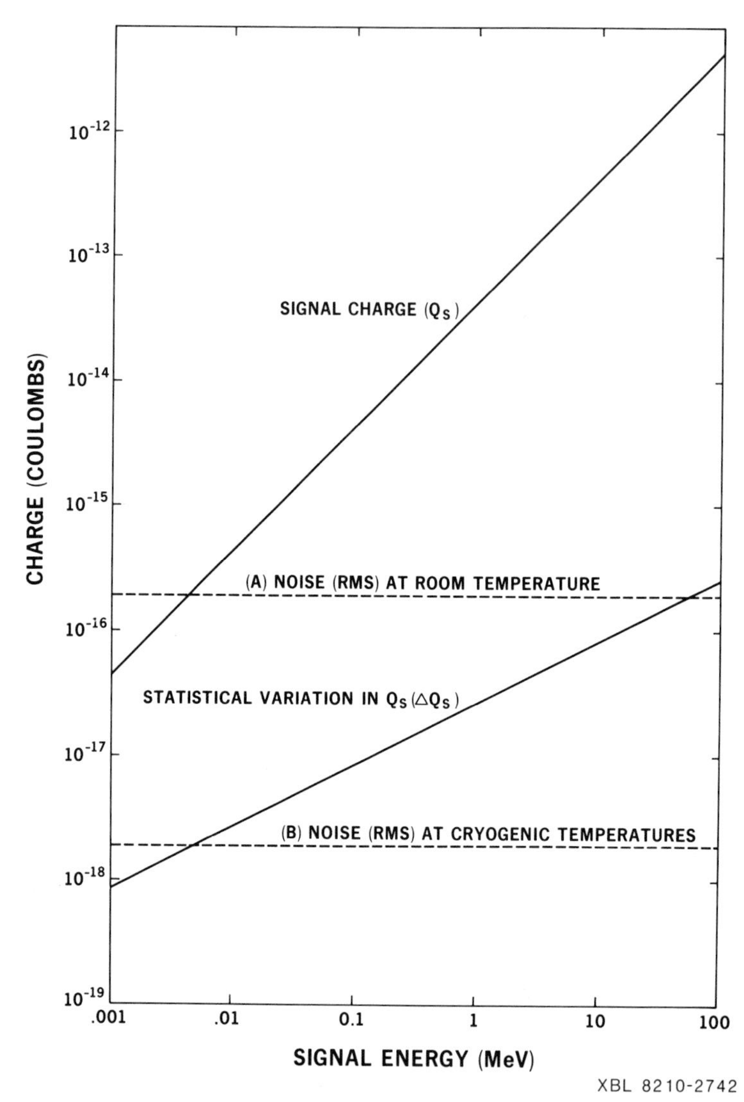

XBL 8210-2742

Fig. 3. Signal charge, Q_S, (coulombs) produced by incident signal radiation (MeV) and the statistical variation, ΔQ_S, in this signal charge for silicon nuclear radiation detectors. The noise of a representative detector and electronics at room temperature (A) and at cryogenic temperatures (B) are also shown.

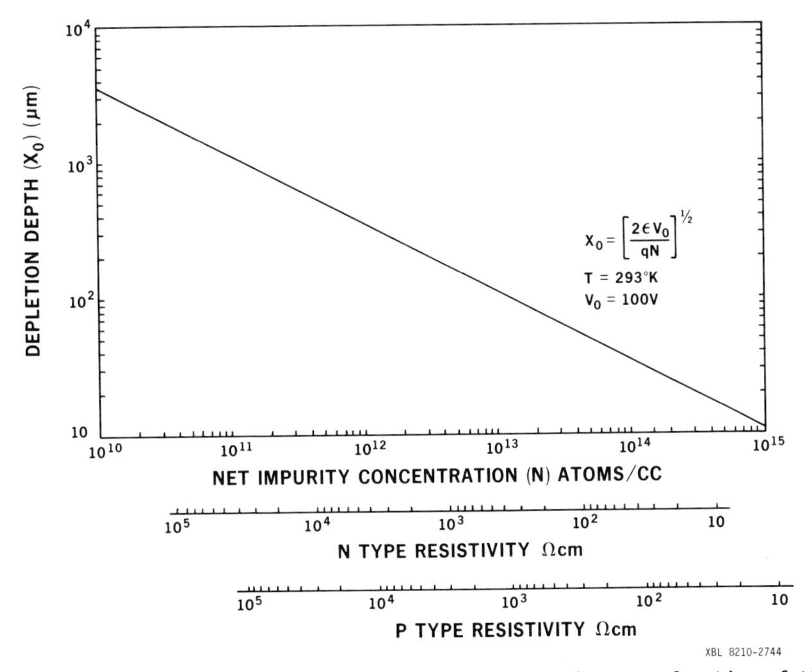

Fig. 5. Depletion width (μm) with 100 volts reverse bias as a function of the net impurity concentrations (atoms/cc). Also shown are the n-type and p-type crystals' resistivities corresponding to the impurity concentration.

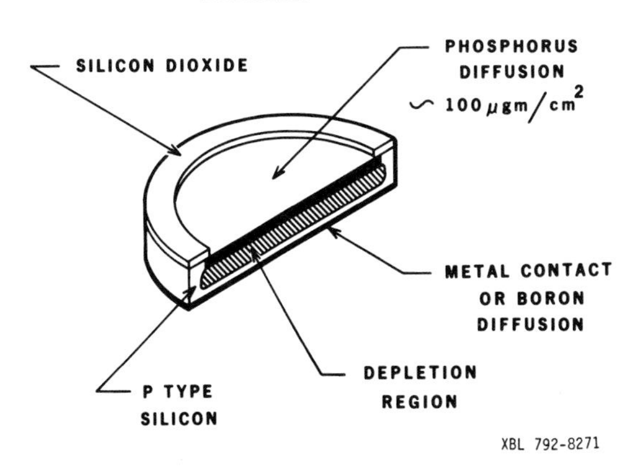

Fig. 6. Schematic of a diffused junction detector.

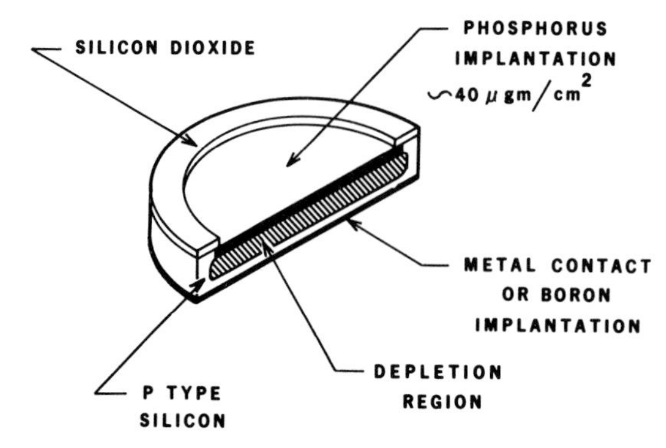

Fig. 7. Schematic of an ion-implanted junction detector.

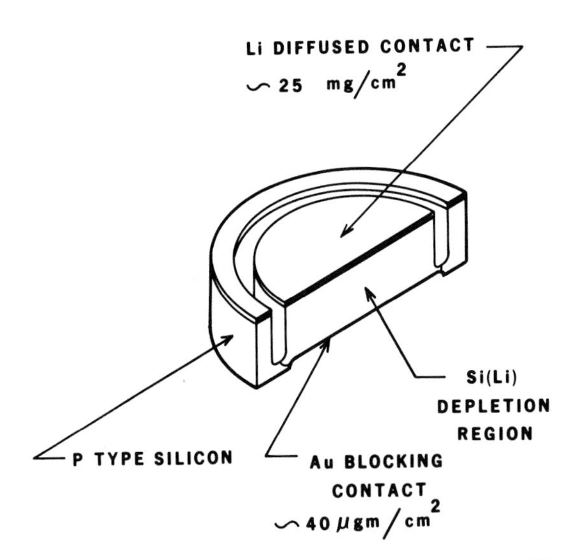

Fig. 8. Schematic of a lithium-ion compensated detector.

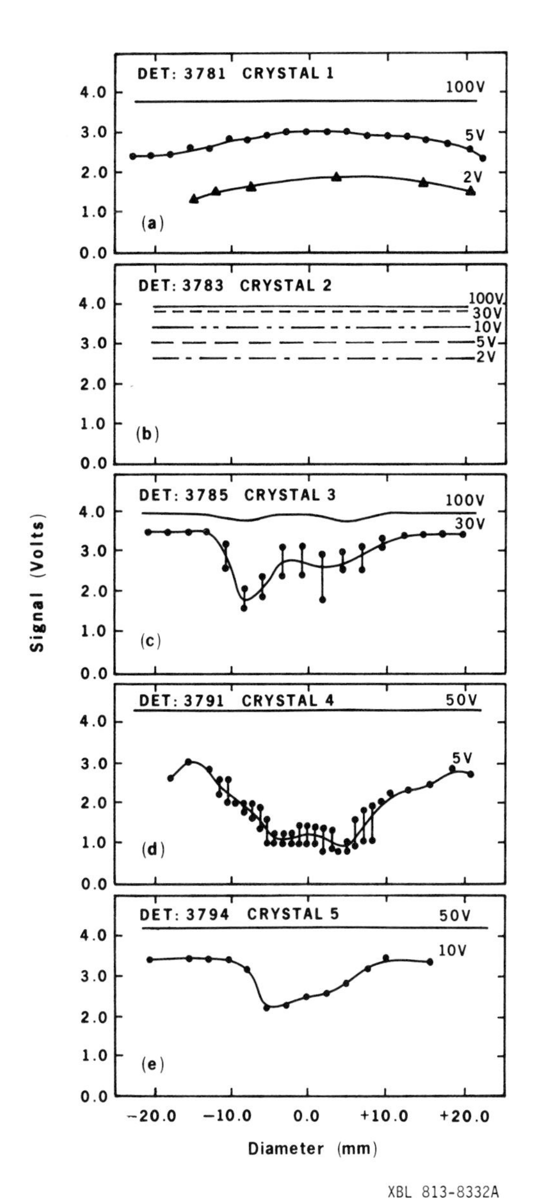

XBL 813-8332A

Fig. 9. The alpha particle scans of detectors made from five different crystals with various applied voltages. The lines connecting points in (c) and (d) indicate the presence of multiple peaks.

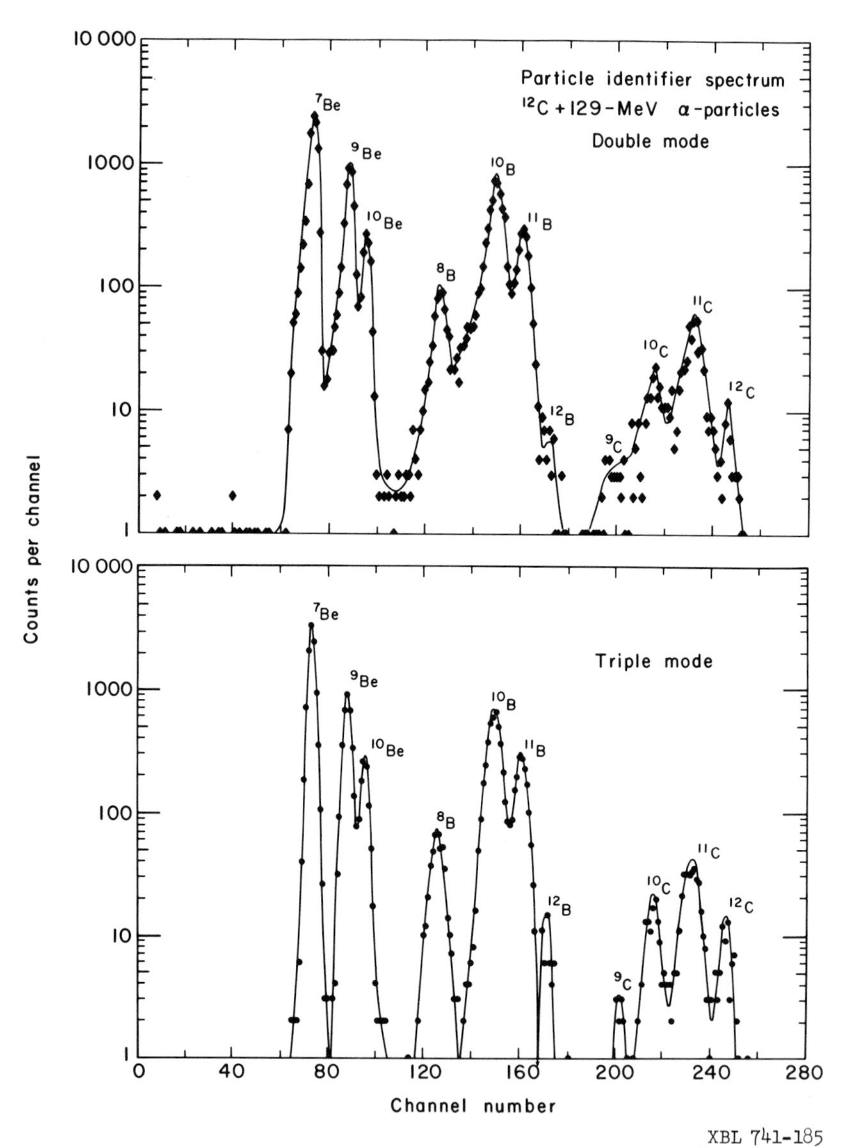

XBL 741-185

Fig. 10. The particle identification spectrum produced by (a) a dual element ΔE, E telescope and (b) a triple element ΔE,ΔE,E telescope on the same ^{12}C + 129MeV alpha experiment.

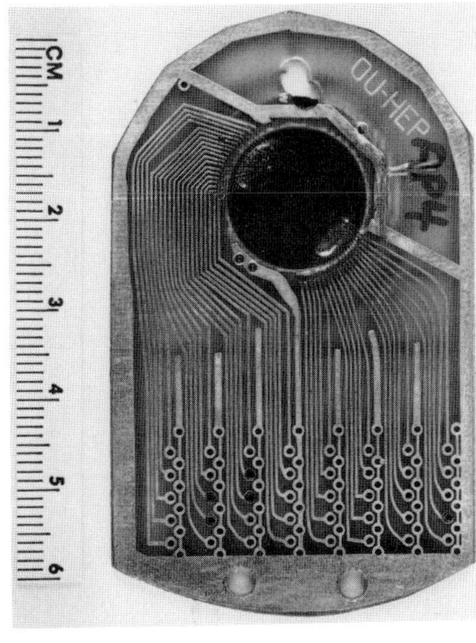

Fig. 11. Prototype vertex detector with 48 stripes, 1cm long, 20μm wide and 60μm apart. Only 40 of the stripes are connected by aluminum wires which are potted in epoxy to improve the device ruggedness.

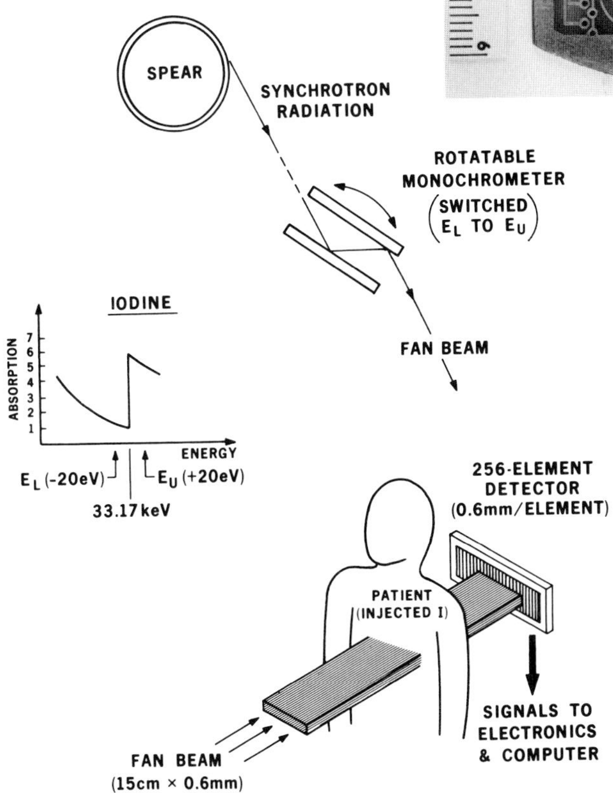

Fig. 12. Pictorial representation of the system for performing angiography using synchrotron radiation.

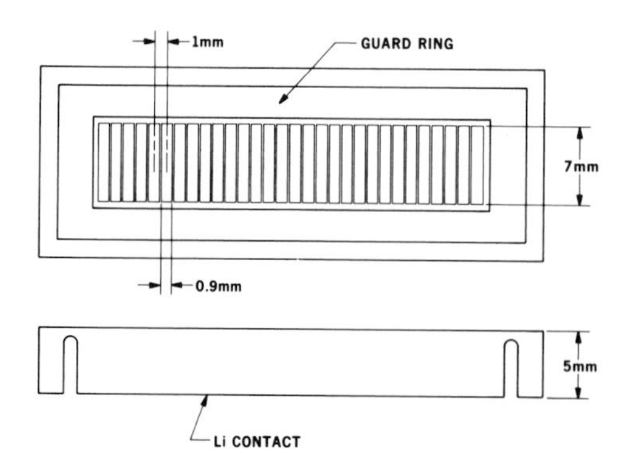

GEOMETRY

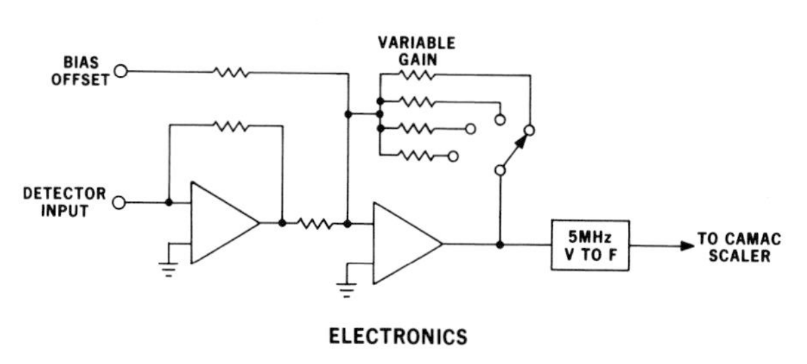

MULTIELEMENT Si(Li) DETECTOR

XBL 817-10670

Fig. 13. Geometry of the 30-element silicon lithium-ion compensated detector used to produce the images in Fig. 14.

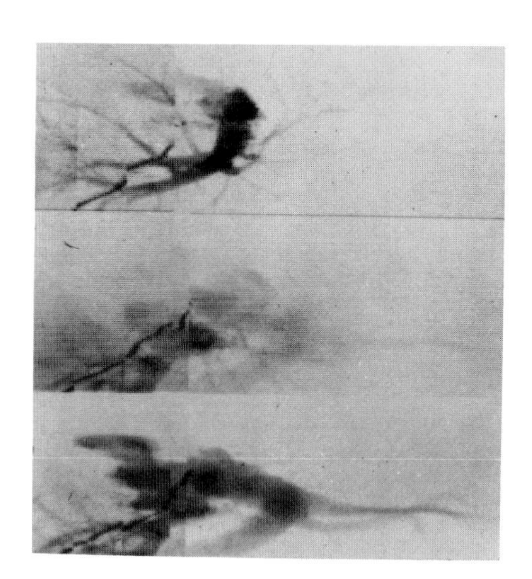

Fig. 14. Angiographs of blood flow in a dog's heart. The pictures are at two-second intervals after the injection of iodine with a catheter (top frame).

MATERIALS ASPECTS OF GERMANIUM RADIATION DETECTOR FABRICATION

G. SCOTT HUBBARD
Canberra Semiconductor, Inc., 24 Digital Drive, Novato, CA 94947

ABSTRACT

An overview of the present status and future requirements of material for germanium radiation detectors is presented. Fabrication and storage problems for both lithium-drifted, doped germanium and high-purity germanium are compared to demonstrate the reasons for the recent complete dominance of the latter in commercially available radiation detectors. The effect of electrically active point and line defects on the resolution and operating characteristic of high-purity germanium radiation detectors is discussed. Emphasis is placed on deep impurities and dislocations. Present and future applications of high-purity germanium radiation detectors are reviewed.

INTRODUCTION

During the past dozen years a major change has taken place in the material used for germanium radiation detectors. From 1962 [1] to 1970 all such detectors were produced utilizing the lithium-drifting technique developed by Pell in 1960 [2]. By compensating p-type germanium containing 10^{13} to 10^{14} cm^{-3} impurities with an interstitial donor (lithium) it became possible to obtain extremely low net impurity concentrations. Concentrations of less than 10^{9} cm^{-3} have been routinely reached. Using this method, lithium-drifted Ge(Li) devices with an active volume of 150 cm^{3} or more were eventually successfully fabricated. This result becomes all the more impressive when compared to the first Ge(Li) detector with a sensitive region of .2cm^{3}. However, shortcomings in the material (detailed in Table I) combined with a proposal for high-purity germanium (hpGe) put forth by R.N.Hall in 1965 [3] led to efforts by several laboratories to produce inherently pure crystals. In 1971 Hall [4] and Hansen [5] demonstrated the feasibility of growing monocrystalline germanium with net impurity concentrations of $\leq 10^{10}$ cm^{-3} (as measured at 77K). Significant efforts at commercial exploitation followed 4-5 years later. Today no detector manufacturer of which the author is aware is producing any significant quantities of lithium-drifted germanium radiation detectors. This changeover was propelled by two very practical factors: (1) the need to always keep Ge(Li) detectors at cryogenic temperatures (or risk decompensation) and (2) the long drifting periods (up to 100 days) compared with high-purity germanium detector fabrication times of a few days including testing. As an added benefit availability of this new high-purity material has resulted in many applications which would have been extremely difficult or impossible before.

This review will begin with a brief summary of the general materials properties of germanium radiation detectors, then concern itself with current materials problems in high-purity detector fabrication and conclude with recent and future applications unique to hpGe.

TABLE I
A Comparison of Lithium-drifted and High-Purity Germanium Coaxial Detectors

	Ge(Li)	hpGe
Starting Material		
Cost	~$5 per gram	~$10 per gram
Level of Technology	Mature, high volume	Still partly under development, occasional problems in production for all manufacturers
Detectors		
Fabrication and Testing Time (assuming no unusual complications)	>60 days drift time for $60 cm^3$ active volume	5-10 days
Thermal Cyclability (from room temperatures to cryogenic temperatures and back)	Detector must be cold at all times to prevent decompensation	Essentially unlimited thermal cycles
Detector Geometry	n^+-contact outside p^+-contact in the center	Contact may be reversed depending on need
Charge Collection	Many times undepleted material present resulting in "slow" charge pulses which travel only by diffusion	Rarely undepleted material
^{60}Co Resolution (FWHM) for state of the art devices of equal size (E_γ=1.33 MeV line)	$\leq$2.0keV	$\leq$2.0 keV
Peak to Compton ratio	May be reduced by undepleted material	Improved by lack of undepleted material
Fast Neutron Damage	Repaired only by redrifting	May be thermally annealed, sometimes in situ.

I. The Germanium Radiation Detector

High resolution spectroscopy of energetic photon and charged particle radiation is required in many industrial and research applications. The semiconductor radiation detector, first developed to study nuclear and atomic structure, is by far the best tool for such work. A sodium iodide NaI(Tl)-photomultiplier tube combination (the so-called scintillation detector) offers the most comparable alternative to semiconductor radiation detectors. However, Figure 1 demonstrates that where high resolution is required, sodium iodide cannot compete effectively.

A semiconductor detector is an p^+-i-n^+-diode operated under reverse bias sufficient to achieve good charge collection throughout the entire volume. At the same time leakage current (I_L) must be low enough not to introduce significant noise into the system. For most applications I_L must be well below 10^{-9}A. For photons in the range of approximately 30 keV to several MeV, the semiconductor of choice is germanium. Material properties which dictate this selection are as follows:

-- Only 3.0 eV is required to produce an electron-hole pair in germanium. By comparison, for silicon 3.8 eV are necessary (at 77K). Furthermore, the charge produced in either material is directly proportional to the energy of the incoming radiation:

$$N = E \text{ photon (eV)} / \varepsilon \tag{1}$$

where N is the number of photon hole-electron pairs and ε is the energy required to produce a hole-electron pair.

-- Carrier mobility ($\mu_{e,h}$) in germanium is $4.0\text{x}10^4 \text{cm}^2/\text{Vs}$ at 77K for both holes and detectors. [The diodes must be operated at cryogenic temperatures to avoid the high leakage current associated with a small band gap material (.7eV) in room temperature operation]. By contrast $\mu_e = 1.9\text{x}10^4$ and $\mu_h = 3\text{x}10^3$ for Si at 77K. At an electric field of ~1000 V/cm in the bulk of the germanium device a charge carrier will then move at a saturation drift velocity of 10^7 cm/sec.

-- The probability with which an incoming photon will interact with an electron in a material and transfer all its energy into the kinetic energy of the electron (photoelectric effect) has a Z^5 dependence on the atomic number of the material. Consequently, germanium is again the material of choice above about 30 keV (provided the sensitive volume is of sufficient length.) The greater density of germanium also makes it the principal detector material for charged particles of energy up to that of 200 MeV protons, or equivalent.

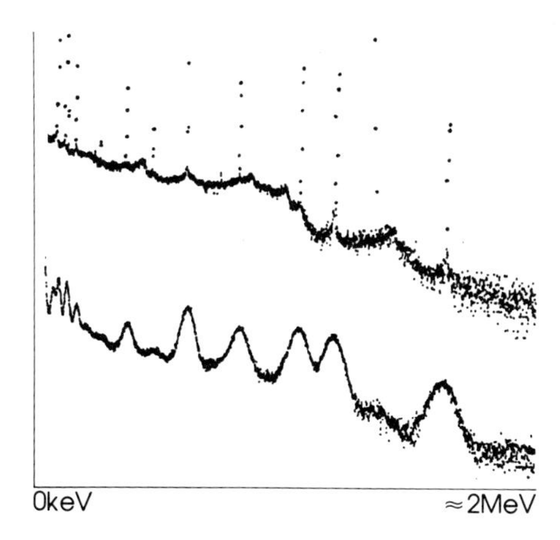

Figure 1. Comparison of a 3"x3" NaI(Tl) and a Ge(Li) radiation detector using a NBS Mixed Point Source [SRM 4215C]. Resolution of the Ge(Li) detector is 2.0 keV (FWHM) at the ^{60}Co 1.332 MeV line. Resolution of the NaI(Tl) detector is 7½% at ^{137}Cs .662 MeV line, 100 sec count time.

Much more detailed descriptions of the physics of the semiconductor detector and charge production processes are available, particularly the recent reviews by Haller, et al [6,7]. For our purposes, it is sufficient to understand the operation of the so-called planar radiation detector shown in Figure 2. Under sufficient reverse bias a volume depleted of free charge carriers is formed. For optimum charge collection the electric field should be 1000V / cm in the entire bulk. When incoming radiation is absorbed, electrons and holes are produced which move at their drift velocity to the appropriate contact. There they are collected by the first stage of a charge sensitive preamplifier. Subsequent amplification and pulse height analysis adds the pulse to the accumulated histogram which eventually becomes the characteristic spectrum of the source.

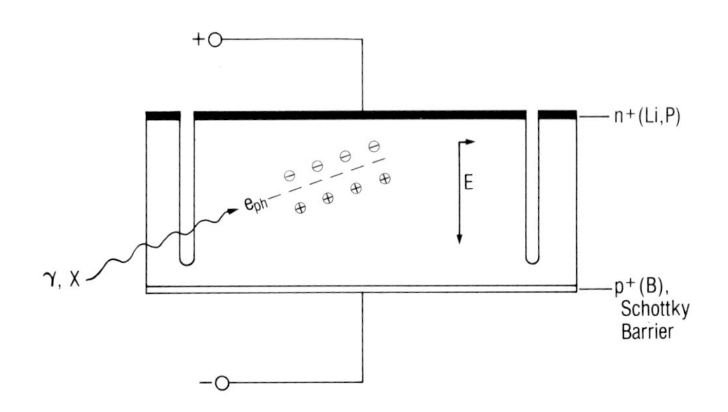

Figure 2. High-purity germanium (hpGe) radiation detector. Grooved structure is used to define the active area and facilitate handling. Typical active area diameter for many applications is 35mm or less, while the usual thickness ranges from 5-15 mm.

While the planar structure lends itself to the analysis and discussion of the physics of the device the coaxial detector is more widely used because of its greater volume and therefore greater efficiency * with weak sources. Figure 3 shows the two basic geometries currently in widest use. It is possible to fabricate a "true" coaxial detector by drilling the hole completely through the crystal but then the detector maker is faced with protecting two sensitive surfaces rather than one. Principally for this reason, commercial manufacturers avoid this geometry. With a closed coaxial detector, the only variation is the orientation of the contacts. We will compare these two possibilities in section IV. Regardless of geometry, the detector maker is always faced with the same broad classes of tasks: --material selection and availability; --contact formation; --surface passivation. We shall defer the first problem until section three and take up the other two now.

II. Detector Fabrication

Even a casual survey of papers written during the last ten years on the subject of producing radiation detectors from hpGe reveals that many different techniques will work for some people at least some of the time. A summary of

---

* By convention, the active volume of commercially available coaxial detectors is measured by comparison to the efficiency of 3"x3" NaI(Tl) detector for a standard γ-ray source.

contact fabrication and surface treatments which appear to be commonly used is shown in Table II. No attempt at completeness has been made since the published surface etching recipes would fill several pages alone. Given that the ultimate performance is equivalent, the choice of which method to use is then usually dictated by other considerations such as cost, simplicity, usability for a production process, etc.

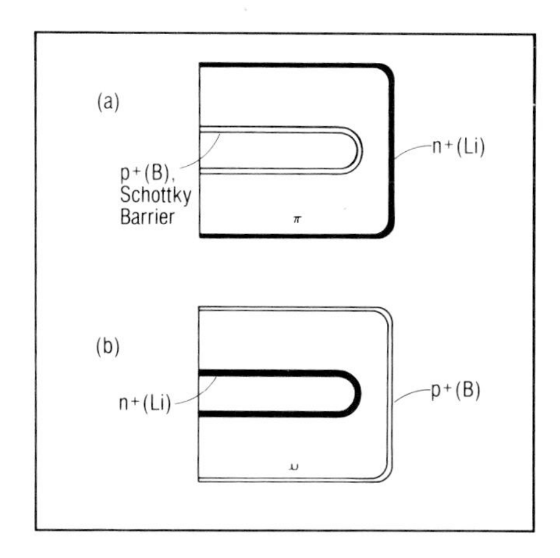

Figure 3. Coaxial structure hpGe radiation detectors. Device (a) has standard electrode geometry (SEGe); detector (b) has been fabricated with reverse electrode geometry (REGe). In principle the type of germanium is of little concern but to avoid extremely high fields at the core contact SEGe detectors are usually made from p-type and REGe from n-type germanium.

It does appear that the extreme reproducibility of boron implantation for p^+-contacts may outweigh the cost and complexity of the ion implantation equipment and become the method of choice. It would also seem that the inherent problems associated with wet chemistry may lead to the dominance of some "dry" technique like sputtered germanium for final surface passivation.

A continuous source of problems for the detector manufacturer is the inability to easily distinguish between fabrication technology and the germanium crystal as a cause of poor device performance. In an ideal detector, the leakage current (I_L) vs applied reverse bias (V) characteristic should show $I_L \approx 10^{-11}$ A at V=5000 volts. If the contribution of electronic noise is negligible and the Fano factor [16] is ~0.1, the resolution of the ^{60}Co γ-ray photopeak at 1.33 MeV should be 1.7 keV as measured by its full width at half-maximum (FWHM). In the absence of any charge trapping mechanism the full width at tenth maximum (FWTM) should be ~3.1 keV. In addition efficiency measurements should show no "dead" or noncollecting material.

Under realistic conditions the performance of the ideal detector can only be approached. At times a detector might show a "resistive" I_L vs V characteristic. This means the leakage current rises in proportion to the reverse bias. As a consequence the electronic noise usually increases substantially and the ^{60}Co photopeak will broaden significantly. The resistive I_L-V characteristic may be caused by an uncontrolled or poorly executed step in the fabrication. Two examples are: contacts which inject rather than rectify and surface leakage currents. A persistant and careful detector maker who repetitively experiences problems with a given crystal may eventually conclude

TABLE II

Contact Formation for hpGe Radiation Detectors

	Advantages	Disadvantages	Typical Technique
n^+-Contacts			
Lithium evaporation and diffusion [8]	Conditions well controlled, only elemental Li present	Vacuum required Coverage difficult for coaxial detectors Up to few $X10^2$ μm thick	Evaporate Li for 3 mins under 10^{-6}T vacuum. Backfill with Argon to $4x10^{-2}$T diffuse 7 mins at 280°C
Lithium electroplating and diffusion [9]	Apparatus needed very simple, easy to use	Hot Li-salts hazardous Extensive masking with Teflon tape required	Protect Ge with teflon tape, submerge device in Li salt bath, plate, diffuse 10 mins
Phosphorous Implantation [10]	Ultra-thin contact (~1000Å) no "dead" material May be annealed without increasing dead layer	Complicated process Not as reproducible or as rugged as Lithium Implantation facility is required	Implant 10^{15} cm^{-2} ^{31}P at ~77K. Pre-anneal @ 150°C for 24 hours Anneal @ 330°C for one hour
p^+-Contacts or Equivalent			
Boron implantation [11]	Extremely reproducible contact Only practical way to make REGe detector	Implantation facility required	Implant 10^{14} cm^{-2} $^{11}B+$ ions into etched surface
Schottky-Barrier Thin Metal Film evaporation (Au, Pd most common) [12]	Equipment readily available Mature, relatively simple technology	Success of barrier depends on preparation of surface Spalling possible	Use hot filament or e-gun to evaporate 1000Å of metal onto etched surface under 10^{-6}to10^{-7}T vacuum
Surface Treatment			
Wet Chemistry [13]	Inexpensive Easy to use Necessary step for removing saw or lapping damage	Uncontrolled particularly during quenching Etching rate depends strongly on temperature of chemicals	Protect contacts, etch 1-2 mins in a mixture of 3:1 [HNO_3:HF], quench with methanol, dry with N2
SiO Evaporation [14]	Claimed to protect and stabilize surface states	Requires vacuum technology Depends on "art" of operation	Evaporate 500-1000Å of SiO on etched surface
α-GE [15] (sputtered amorphous germanium)	Very controllable and apparently quite reproducible Passivates and protects sensitive surface	Requires RF Sputtering equipment. Much more complicated than above techniques	RF sputter 3000Å of amorphous, hydrogenated germanium on etched surface

that it is not the processing but the material which is at fault. As we shall see, the cause for materials related problems may be residual electrically active defects.

III. Current Problems in hpGe Production

A growing number of laboratories are successfully producing single crystal germanium which fulfills the basic requirements of purity, crystallography and size for detector grade germanium. [17] As this development has progressed, attention has shifted from these prerequisites to the problems of improving charge collection and determining the link between material properties and detector characteristics. [18] Currently all but the largest detectors must have a ^{60}Co resolution of 2 keV at 1.33 MeV in order to be considered state of the art and commercially competitive. In order to achieve this resolution, the crystal must be free of charge-trapping centers, particularly deep levels in the energy range of ~30 meV to ~200 meV where the trapping - detrapping times most seriously affect resolution. Several authors have reported copper, copper impurity-complexes and dislocations to be the most common sources of deep traps in this energy range. [19,20] As an example, Figure 4 compares the Deep Level Transient Spectroscopy (DLTS)* data for two diodes of equivalent size, operating bias and impurity concentration fabricated from crystals grown at Canberra Semiconductor (CSI). SEGe coaxial detectors were made from the crystal section adjacent to the DLTS samples. Detector (a) exhibited very poor resolution (^{60}Co FWHM=3.5 keV) due to the presence of copper and copper complexes. Note that the ratio of Cu_s(44) to CuH(175) is ~2:1. By comparison the ^{60}Co FWHM of detector (b) = 1.7 keV even though copper-hydrogen complexes are still very much in evidence. Capacitance and Hall effect measurements indicate that the

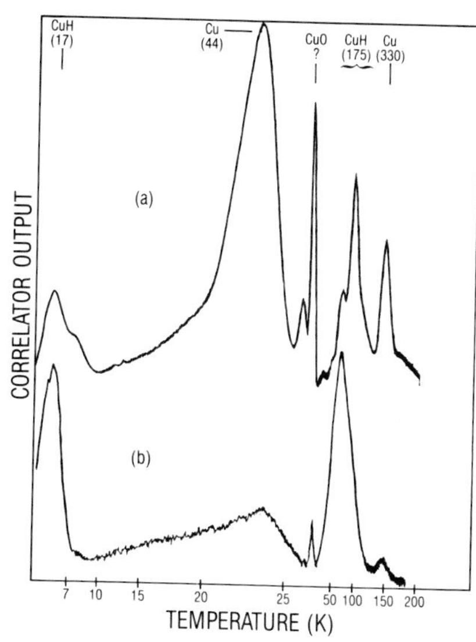

Figure 4. DLT spectrum of two diodes containing copper and copper-hydrogen acceptors. Numbers in parenthesis indicate activation energies in meV. Ten per cent (10%) efficiency SEGe coaxial radiation detectors made from sections adjacent to these diodes exhibited ^{60}Co FWHM resolution at 1.33 MeV of 3.5 keV (a) and 1.7 keV (b).

---

* DLTS is now a common technique in semiconductor materials analysis. See references [21,22] for a more detailed description.

total concentration of deep levels of the two samples is comparable but the ratio of Cu:CuH in sample (b) is much in favor of CuH. This result implies that both the energy distribution of deep levels and their concentrations are important. In particular, one group has determined that a concentration of substitutional copper $Cu_S(44) \sim 5 \times 10^9 cm^{-3}$ will seriously affect detector resolution. [23]

The effect of dislocations on device performance is of great importance to the whole field of semiconductor development as well as radiation detectors. Unlike silicon crystals employed by integrated circuit manufacturers, dislocation-free germanium cannot be used for radiation detectors. A divacancy-hydrogen complex (V_2H) which has an activation energy of E_v+.08 eV is a severe charge trap rendering dislocation-free hpGe useless for detectors. [24] Glasow and Haller have demonstrated the correlation between dislocation density $\sim 10^4 cm^{-2}$ (as determined by etch pit counting) and poor ^{60}Co resolution in hpGe planar detectors. This effect is due to bands of acceptor levels which appear at dislocation densities of $5\text{-}10 \times 10^3 cm^{-2}$. Figure 5 (from Hubbard and Haller [25]) shows the DLTS spectra of hpGe samples with increasing defect densities. As the etch pits begin to cluster the acceptor levels appear. Figure 6 shows a DLTS spectrum from a CSI sample which was adjacent to a 10% efficient SEGe coaxial detector. Counting etch pits on the [100] plane revealed an etch pit density (EPD) of $3 \times 10^3 cm^{-2}$ in the center and $8 \times 10^3 cm^{-2}$ at the edge. The device had a ^{60}Co (FWHM) resolution of 2.3 keV when operated at 3000 V. (Depletion voltage = 2000 V.) As shown in Figure 6 the I_L-V characteristic was resistive at this point. We then removed approximately 20% of the mass of the device from the periphery (the volume containing the high EPD) and found the resolution improved to 2.1 keV at 3000 V. In both cases the electronic noise contribution was less than 1 keV. In addition the I_L-V curve in Figure 7 shows a much better reverse diode characteristic. Both the resolution and I_L-V characteristic were reproduced through repeated surface treatments implying that

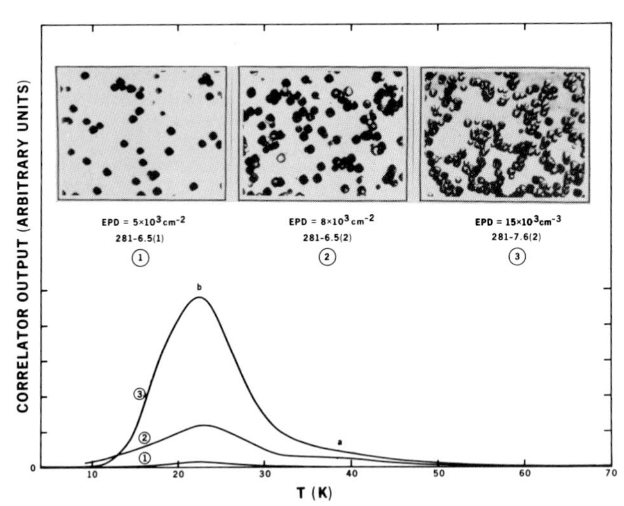

Figure 5. DLT spectrum of defect states associated with increasing dislocation density.

the high EPD was responsible for both the degradation in charge collection and operating characteristic. While the improvement in resolution was not dramatic the result indicates the sort of problems which become very important in larger devices where the charge collection distances are longer.

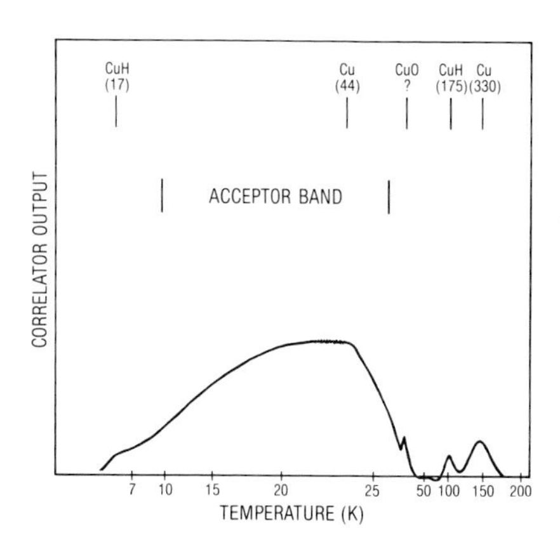

Figure 6. DTL spectrum of a diode showing the acceptor band characteristic of high dislocation density. EPD=$8x10^3 cm^{-2}$ near the edge of the device.

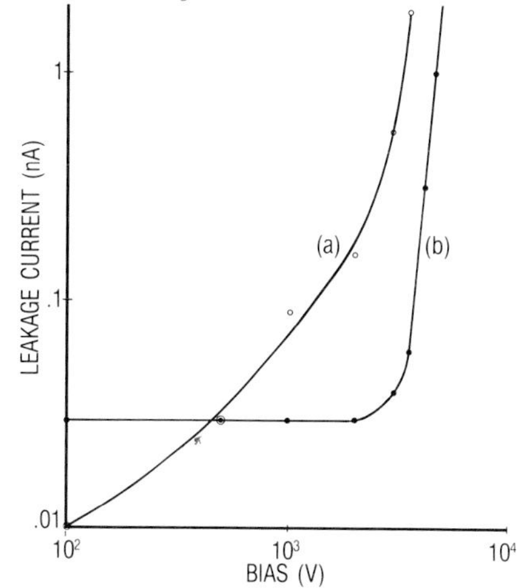

Figure 7. I_L vs V characteristic for SEGe coaxial detector with high EPD at edge (a) and with high EPD volume removed (b).

IV. Applications

With the ascendency of hpGe as the material of choice for germanium radiation detectors many routine measurement procedures have become more efficient and new applications have become possible which were not conceivable before. Several typical examples are given below: for a more complete listing the reader should refer to the following papers [26,27].

High-purity germanium has provided charged particle spectroscopy a powerful new tool: the detector telescope. [28] Figure 8 shows a stack of planar detectors (7 Ge, 1 Si(Li)) currently in use at the Los Alamos Meson Physics Facility (LAMPF). Each germanium detector has two ultra-thin contacts made by implantation of $^{11}B+$ and $^{31}P+$. By eliminating the undesirable dead layer due to thick lithium diffused contact, experimenters have been able to gather data formerly obtainable only with the magnetic spectrometer which is much larger, more expensive and covers only a limited energy range at one time. The proton peak resolution in Figure 9 from an experiment at Indiana University may be accounted for by beam spread, reaction kinematics, detector statistics and electronic noise. The contribution of the detector must clearly be quite small.

Figure 10 shows the schematic of a hand-held Multi-Attitude Cryostat (MAC) developed by Canberra Industries for use in the Darmstadt-Heidelberg Crystal Ball [29]. As seen in the figure, the apparatus is constructed to accept a large number of detector modules in a 4π geometry. Originally constructed using only NaI(Tl) detectors, germanium detectors were added to perform multiple anti-coincidence experiments. The operating requirements were that the spectrometer be able to function in any position without consuming the space usually devoted to the standard 31 liter dewar and consequently be repeatedly thermally cycled without difficulty. Since Ge(Li) detectors must be kept cold continously to avoid decompensation only hpGe can be used in compact designs.

Figure 8. An eight-detector fixed position telescope (7 high-purity germanium and 1 Si(Li)) is shown. The telescope is one of two being used in coincidence experiments at LAMPF. Each germanium device has a ^{31}P-implanted n^+-contact. Total germanium thickness is 83 mm.

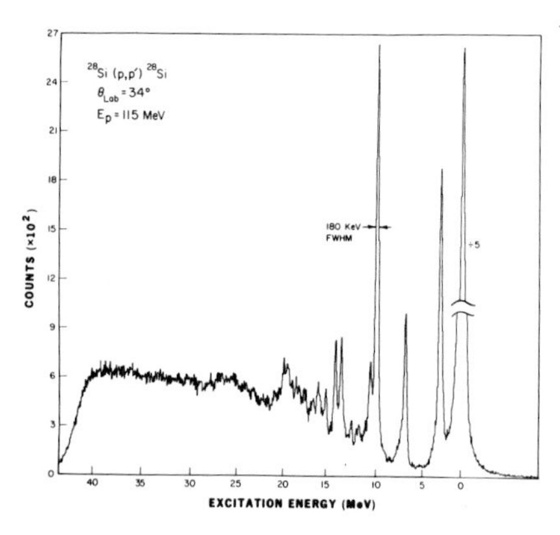

Figure 9. Energy spectrum for the scattering of 115 MeV protons from a 10 mg/cm^2 ^{28}Si target. The spectrum was observed with a detector telescope consisting of two 15 mm thick high-purity germanium detectors with ultra-thin contacts. The proton beam, accelerated in the Indiana University Cyclotron, had a resolution of about 110 keV.

The greatest single impact in terms of both economics and spectroscopy has been due to the development of the reverse electrode coaxial radiation detector. [30] As indicated earlier this device is usually fabricated using n-type hpGe with an ion implanted P^+-contact on the outside and a diffused lithium contact

Figure 10. At the left is the superstructure of the Crystal Ball, capable of holding a 4π array of NaI(Tl) detectors. Above is the hpGe MAC which can be inserted in any position.

in the hole. In addition to the very thin outer contact, which makes possible low energy x-ray spectroscopy, trapping caused by fast neutron damage does not appreciably degrade the resolution until the total neutron dose reaches 20 or 30 times that for SEGe coaxes. [31] In germanium, neutron damage results in only hole traps. Since the greatest detector volume is near the outer contact, reversing the contact causes holes to be collected over the shortest possible distance, minimizing the effect of the damage. Furthermore damage can be easily repaired by thermal annealing rather than requiring redrifting.

Proposed future applications for hpGe radiation detectors include many which will test the limits and resourcefulness of material producers and detector manufacturers. In nuclear medicine and environmental monitoring for example, large arrays of planar detectors or single devices divided into many segments have been considered or tested by prototype. [32] Some experimenters have conjectured that crystals of up to 8cm diameter might be used in such a program. Currently the largest diameter available is ~6 cm. Such an increase in size would mean a geometrical increase in the requirements for purity and trap-free germanium. It would be both challenging and expensive to advance to such large diameters.

Neutron activation analysis of geologic strata in a bore-hole several thousand feet in depth has been successfully carried out using NaI(Tl) and to some extent hpGe. [33] The neutron resistant qualities of REGe detectors would be ideal in this operation. Since one is restricted by the size of the bore hole the ultimate goal would be to grow crystals and produce REGe detectors which are quite long in order to achieve greater efficiency. A detector 4.5 cm in diameter by 6-8 cm long would be most useful in this application. Such a device would require extraordinary control of the crystal purity over a considerable distance in order to achieve approximately equal charge collection in the entire volume.

In nuclear physics research there exists great interest in the measurement of the mass of the neutrino by means of double beta-decay. With the advent of hpGe and improvements in fast, low noise, pulse electronics, such an experiment is now possible. It will require, however, an array of eight coaxial detectors each 5.0 - 5.5 cm diameter and 7 - 8 cm long. [34] Material of this size is not at present readily available. To consistently produce such volumes of germanium will require an advance in the state of the art of crystal growing as well as detector fabrication.

V. Conclusion

We have seen that advances in the purification and growth of germanium crystals have led to the development and commercial exploitation of high-purity germanium for radiation detectors. As a consequence no commercial supplier is currently producing Ge(Li) radiation detectors. With substantial quantities of hpGe becoming available attention in both the materials research and industrial sectors is being focussed on improvements in device resolution and operating characteristic. A principal area where such progress can be made is in understanding and controlling point and line defects which result in deep trapping levels. Two common examples are copper related species and dislocations. Proposed projects requiring large volume coaxial detectors, detector arrays or large area planar detectors lead to very long charge collection distances which demand that charge trapping centers must be reduced to absolute minimums. Carrying out this development will require that the investigations and analysis done by materials researchers be extended and that the cooperation already established between Universities, National Laboratories, and Industry be increased.

Acknowledgements

The author wishes to thank E.E. Haller and W.L. Hansen for providing the DLTS measurements on CSI crystals as well as constructive discussions on many occasions. O. Tench and staff at Canberra Industries have made available all the data on the performance of the detectors mentioned. Finally, I wish to thank G. Mraz, J. Furst and J. Gove for growing crystals, making measurements and helping to prepare this manuscript.

References

1. D.W. Freck, J. Wakefield, Nature 4816, 669 (1962).
2. E.M. Pell, J. Appl. Phys. 31, 291 (1960).
3. R.N. Hall, "Semiconductor Materials for Gamma Ray Detectors," p. 27, Proc. of the meeting June 24, 1966 N.Y. N.Y., eds. W.L. Brown (BTL) and S. Wagner (BNL).
4. R.N. Hall and T.J. Soltys, IEEE Trans. Nucl. Sci. NS-18, 160 (1971).
5. W.L. Hansen, Nucl. Instr. Meth. 94, 377 (1971).
6. E.E. Haller, IEEE Trans. Nucl. Sci. Vol NS-29, No. 3, June 1982.
7. E.E. Haller and F.S. Goulding, Handbook on Semiconductors, Vol. 4, Ch. 6C, ed. C. Hilsum, North-Holland Publ. Co. (1980).
8. See for example: R.H. Pehl, R.C. Cordi and F.S. Goulding, IEEE Trans. Nucl. Sci. NS-19, No. 1, 265 (Feb., 1972).
9. W.L. Hansen and E.E. Haller, IEEE Trans. Nucl. Sci., Vol. NS-28, No. 1, 541 (1981).
10. G.S. Hubbard, E.E. Haller and W.L. Hansen, IEEE Trans. on Nucl. Sci., NS-24, No. 1, 161 (February 1977).
11. J.P. Ponpon, J.J. Grob, R. Stuck, P. Burger and P. Siffert, Proc. of the II Int. Conf. on Ion Implantation - Semiconductors, Springer-Verlag, New York (1971).
12. For a complete survey see: Handbook of Thin Film Technology, ed. L.I, Maissel, R. Glang, McGraw-Hill (1970).
13. R.D. Baertsch, IEEE Trans. NS-21, 347 (1974).
14. R.J. Dinger, J. Electrochem. Soc. 123, 1398 (1976) and R.J. Dinger, IEEE Trans. NS-22, 135 (1975).
15. W.L. Hansen, E.E. Haller and G.S. Hubbard, IEEE Trans. Nucl. Sci. NS-27, 247 (1980).

16. U. Fano, Phys. Rev. 70, 44,(1946).
17. Lawrence Berkeley Laboratory, Univ. of California, Berkeley, CA, USA. General Electric Company, King of Prussia, Pennsylvania, USA. Metallurgie Hoboken-Overpelt, Division of Chemical Products, Olen, Belgium. EG&G, Oak Ridge, Tennessee, USA. Canberra Semiconductor, Novato, California, USA.
18. E.E. Haller, P.P. Li, G.S. Hubbard and W.L. Hansen, IEEE Trans. Nucl. Sci. NS-26, 265, (1979).
19. A.O. Evwaraye, R.N. Hall and T.J. Soltys, IEEE Trans. Nucl. Sci., NS-26, 271 (1979).
20. W.K.H. Schoenmaekers, P. Clauws, K. Van den Steen, J. Broeckx and R. Henck, IEEE Trans. Nucl. Sci., NS-26, 256 (1979).
21. G.L. Miller, D.V. Lang and L.C. Kimerling, Ann. Rev. Mat. Sci., 7, 377 (1977).
22. D.V. Lang, J. Appl. Phys. 45, 3022 (1974).
23. E. Simoen, P. Clauws, J. Broeckx, J. Vennik, M. Van Sande, and L. DeLaet, IEEE Trans. Nucl. Sci., NS-29, No. 1, 789 (1982).
24. E.E. Haller, G.S. Hubbard, W.L. Hansen and A. Seeger, Inst. Phys. Conf. Ser. 31, 309, (1977).
25. G.S. Hubbard and E.E. Haller, J. Electr. Mat. 9, 51 (1980).
26. R.H. Pehl, Physics Today, 30, No. 11, pp 50-61 (November 1977).
27. P.A. Glasow, IEEE Trans. Nucl. Sci., NS-29, No. 3, 1159 (1982).
28. G.S. Hubbard and E.E. Haller, Nuclear Inst. and Methods, 164, 121 (1979).
29. R.S. Simon, Journal De Physique, 41, C10-281 (1980).
30. H.W. Kraner, R.H. Pehl and E.E. Haller, IEEE Trans. on Nucl. Sci., NS-22, No. 1, pp 149-154 (February 1975).
31. R.H. Pehl, N.W. Madden, J.H. Elliott, T.W. Raudorf, R.C. Trammel, and L.S. Darken, Jr., IEEE Trans. Nuc. Sci. NS-26, 321 (1979).
32. J. Patton, R.R. Price, F.D. Rollo, A.B. Brill, R.H. Pehl, IEEE Trans. Nucl. Sci. NS-25, No. 1 (1978).
33. G.J. Schnerr, L.H. Goldman, and P. Ryge, IEEE Trans. Nucl. Sci., NS-27, No. 1, 240 (1980).
34. M.S. Witherell, AIP Proceedings, Workshop on Science Underground, to be published.

THICK SURFACE BARRIER DETECTORS MADE OF ULTRA-HIGH PURITY P-TYPE SI SINGLE CRYSTAL

F. SHIRAISHI, Y. TAKAMI AND M. HOSOE
Rikkyo University, Nagasaka, Yokosuka, Kanagawa, 240-01

Y. OHSAWA AND H. SATO
Komatsu Electric Metals Co., Shinomiya, Hiratsuka, Kanagawa, Japan.

ABSTRACT

The detector material used in this experimments is Ultra-High Purity p-type Si crystal. The material was single-crystalized through floating zone process from poly-crystal grown by thermal decomposition of highly refined mono-silane gas which had been purified by molecular sieves of specially prepared Zeolite powder.

The resisitivity at room temperature is normally above 30 kΩ·cm, and the value of the highest grade ones exceeds 100kΩ·cm which corresponds to the Boron concentration of 1.5×10^{11} B/cm^3.

The potentiality of this material for detector use was investigated through Surface Barrier Detector fabrication. Detectors of above 4 mm thick and of excellent characteristics both at room temperature and at liq. N_2 temperature could readily be fabricated. Owing to the extremely high resisitivity, detectors can be made either partially depleted or totally depleted simply by properly selecting the wafer resisitivity and the thickness.

In detector fabrication, the proper surface chemical treatment is very important, and seriously affects the leakage current and breakdown characteristics.

Ultra-High Purity Si is promising as new detector material and has good potentiality to replace Si(Li) as followings:

1) simple and easy detector fabrication,
2) potentiality of thick detector fabrication ($\sim$1cm),
3) no precipitation problem of Li ions, and
4) feasibility of thick ΔE detector with thin entrance window on both faces.

1. INTRODUCTION

High resisitivity semiconductor material is always favoured to fabricate detectors of thick depletion layers.
The well-known process of Li ion drifting compensates for the acceptor impurities in the p-type Si or Ge and enables it to achieve high resisitivity to provide a thick depletion layer. In Ge detectors, however,a recently developed high purity Ge crystal whose sensitive volume can exceed Ge(Li), is going to replace Ge(Li). On the other hand, thick Si detectors of present use are all made of Si(Li). Then, the driftability of Li ions limits the

Mat. Res. Soc. Symp. Proc. Vol. 16 (1983)

radiation sensitive depth practically up to 6 to 7 mm.
Si(Li) detector fabrication is more complicated and time-consuming in its process than of Surface Barrier Detectors (S.B.D.) and besides Li precipitation on the surface, while being stored, often causes detector characteristics deterioration.

Si poly-crystal is most commonly produced through thermal decomposition of trichloro-silane gas ($HSiCl_3$), which has some substantial disadvantages in preparing extremely high purity Si. The high decomposition temperature and by-products such as HCl or $SiCl_4$ often result from contamination from the vessel wall. Zone refining of the poly-crystal is not very effective to remove B impurity due to the segregation constant which is close to unity. Though high purity p-type Si can be prepared after this raw material undergoes a number of zone refining processes, its utilization as a detector is not very frequent. The Ultra-High Purity p-type Si single crystal (UHP p-si) used in this experiment is produced by an entirely new method from mono-silanegas (SiH_4).In addition to its extremely high resisitivity, the uniformity of the remaining impurity distribution is excellent, thus the possibility to fabricate thick detectors without Li ion drifting is very high, as it has been tried in High Purity Ge detectors. This paper describes the Surface Barrier Detector fabrication and the performance of this new UHP p-Si, paying special attention to the problem of material characteristics deterioration. Thick Si detectors were fabricated to meet various experimental purposes, and their problems associated with these respective usages are also discussed.

2. HIGH PURITY SI SINGLE CRYSTAL

Si crystal produced by the mono-silane method is substantially low in its impurity such as Boron, and contamination from the vessel during thermal decomposition is also low; therefore the thus produced single crystal shows a very high degree of purity compared with the one by the trichloro-silane methnd.
A new method of mono-silane gas purification using specially prepared molecular sieves has been developed(1) and industrialized.
After the molecular sieve adsorption process has been completed, concentration of the main impurity PH_3in mono-silane is lowered down to $\lesssim 0.01$ppb.Si poly-crystal is grown by thermal decomposition of mono-silane gas; and UHP Si single crystal is grown by floating zone technique.
Details of this crystal preparation process are reported elsewhere in this symposium (2). The purified Si single crystal is usually p-type with little remaining impyrity of Boron and the resisitivity exceeds 100 K$\Omega \cdot$ cm ($\sim$ 1.5×10^{11} B/cm^3) at 300°K.

The extremely high resisitivity measurement of this UHP Si was accomplished by a Photoluminescence (P.L.) method (3) which shows directly the concentration of impurities. Fig. 1 shows the results of a P.L. analysis, applied to highly resisitive 80 kΩ·cm samples. Fig. 1-(a) illustrates the P.L.spectra of two UHP Si samples (80 kΩ·cm), and Fig. 1-(b) illustrates compensated Si of equal resisitivity. It is shown that in compensated Si impurity peaks for B, Al, and P are clearly observed, while in UHP p-Si only a trace amount of B is observed.
Poor resisitivity uniformity is a disadvantage of compensated crystals.

Fig. 2 shows the spacial resisitivity distribution of the UHP p-Si wafer which has been used in this experiment. The resisitivity measurement was made by four point probe method. The four point probe method is not very reliable for highly resisitive Si, but a general tendency of the distribution can be demonstrated. Local differences in the resisitivity are within a factor of two. However, it is evident that no severe localization of p- or n-type impurities exist, due to the nature of this crystal which is not a compensated one. The segregation constant of B is close to unity and the main impurity of UHP p-Si is B; these facts lead to the observed good uniformity in the resisitivity distribution. A highly resisitive n-type Si single crystal can also be prepared from UHP p-Si, doping phosphorus, but the resisitivity distribution is not very good. Therefore the resisitivity is limited up to 20 to 30 kΩ·cm at 300°K.

This high resisitivity n-type Si can be used adequately for Si(Au) S.B.D. by a conventional vacuum evaporation technique. However, experimental investigation made for p- and n-type Si shows that UHP p-Si depletes more easily and a larger sensitive depth can be obtained. In this paper, the fabrication method and performance of S.B.D's made of UHP p-Si, whose resisitivity at 300°K is 50 to 110 KΩ-cm, will be described.

3.FABRICATION OF S.B.D'S MADE OF UHP P-SI

Publications dealing with the fabrication of p-type Si S.B.D. are rather few(4)-(10) compared to n-type Si S.B.D.. With p-type Si, conventional etching with a HF- HNO_3 mixed solution leads to large leakage current diodes. And according to our experience, p-type Si is more sensitive to surface contamination - such as minor impurities in de-ionized water. Surface chemical treatment after the wafer etching seriously influences the detector characteristics, especially when the UHP p-Si detector is totally depleted. In UHP p-Si, the depletion layer often extends to the back contact. The surface states of the back contact greatly influence charge collection behaviour due to the carrier generation and injection at the back surface as well as the leakage current characteristics. A good example of the back contact effect is the charge multiplication phenomenon. Charge multiplication is observed when the band structure is inverted near the back contact and the wafer shows an N-P-N structure. Our experimental results of surface chemical treatment effects are as follows:
good reverse current characteristics are observed if the etched wafers are dipped in HF for a few minutes and immersed in either one of the following four solutions:

1. $K_2Cr_2O_7$ solution,(1%, with H_2SO_4 , 15~30 sec.),
2. NH_4OH(28%)+H_2O_2 (30%), (1:1 ,10~30 sec),
3. HNO_3(60%),~1hr.), and
4. de-ionized water, (2~3 days).

For totally depleted detectors, changing the time interval of chemical treatment for the front(Al side) and back(Au side) face, or selecting different chemical surface treatment for the front and the back face resulted in better detector characteristics (11).

For p-type S.B.D.'s the selection of an appropriate evaporation metal is as important as the surface treatment. Theoretically, a good Schottky barrier can be formed with metals of lower work function than that of Si. However, the quantity of detector leakage current does not follow the order of the metal work function. Our experiment shows that Al and Mg produced the best leakage current characteristics. As a normal procedure, Al is vacuum evaporated on the front side as an N^+ contact and Au on the other side as a P^+ contact.

In most cases, no particular protection is used at the fringe of the electrodes, then breakdown voltage characteristics often show a gradual degradation with time. Detectors for ambient temperature use can be protected with epoxy resin, those with protected fringes are more stable in their reverse current characteristics and the degradation with time is less significant.

4. PERFORMANCE OF UHP P-TYPE SI SURFACE BARRIER DETECTORS.

A. Totally Depleted Detectors .

Thick totally depleted S.B.D.'s with thin entrance windows on both faces can easily be fabricated, utilizing UHP p-Si. In Figs. 3 and 4 , pulse height saturation characteristics at room temperature are shown for Alpha injection on both the Au and Al face; the detector is made of 110KΩ·cm p-type Si (24 mm ϕ ,1mmt). At low bias application, multiple peaks are observed because of the inhomogenous distribution of resisitivity or impurities, but at an applied bias of more than 30V (average electric field: 30 V/mm) the effect of inhomogeneity disappears and there is practically no problem in Alpha-ray spectroscopy. The resolutions for Alpha particles are 40 keV and 36 keV for Al and Au face injection in FWHM respectively. When this detector is cooled down to liq. N_2 temperature, the inhomogeneity effect disappears at a lower electric field of 10 V/mm. This fact can be explained by an increase of the carrier mobility during cooling and the resulting decrease of carrier trapping effects. The leakage current of this detector at a bias of 100V and at 23°C, is 0.68 uA; and at 77K the current increase is observed at 80V. The cause of this strange behaviour is not yet understood, but improvement of the breakdown voltage characteristics is achieved by changing the Au electrode surface into a strong p-type. Examples of thicker detectors will be described in the following:

A detector was made of a UHP p-Si wafer of 22 mm in diameter and 12 mm in thickness with a resisitivity of more than 80 KΩ·cm. Electrodes of 10 mm were vacuum evaporated on both faces.(12). Alpha-ray spectra of both the Al and Au face injection at room temperature at 2.8 KV bias are shown in Fig.5. Fig.5 (a) shows a sharp line spectrum of Alpha particles injected to the Al face, while Fig.5 (b) shows a broad peak of Alpha particles to the Au face. Higher bias voltage application to the detector will lead to a total depletion with sufficient energy resolution, but such a high bias application is a severe limitation for the practical use of a thick detector.

A detector of 15.6 mm thickness will be described next. Bi-207 conversion electron spectra injected to the Au and Al face respectively at liq. N_2 temperature are illustrated in Fig.6. The depletion layer extends from the Al side with increasing bias and conversion electrons injected to the Au side are observable at 2.3 KV bias. The pulse amplitude of the conversion electrons injected to the Au face increases with increasing bias, but even at a bias of more than 2.3 KV , it does not reach the saturation value, while the Aℓ side injection pulses are saturated and reach a 100% pulse amplitude. A hole trapping effect in this wafer material is the most probable cause of this saturation characteristics.

An energy spectrum of Cs-137 Gamma-rays measured by this detector at 1 KV and 77K, is shown in Fig.7. The Gamma counting efficiency saturates against an increasing bias of more than 1 KV and a full energy peak of good intensity is observed. The energy resolution of 7 keV FWHM is slightly inferior to thin UHP p-Si S.B.D's.

B. Partially Depleted Detectors.

UHP p-Si S.B.D's with resisitivity of less than 50 KΩ·cm usually partially deplete unless the wafer is very thin, and the detector is not very sensitive to surface chemical treatment. A conversion electron spectrum of Bi-207 at 25°C measured by a partially depleted detector of 1 cm^2 electrode area is shown in Fig. 8. The energy resolution is 30 keV FWHM for 1 MeV electrons and a similar resolution is also obtained for Alpha particles. At 300V bias, the reverse current was 0.25 uA and the measured capacitance corresponded to a 4 mm depletion depth. The detector characteristics had been stable for more than two years without electrode edge protection.

A thick depletion layer is preferred for large area detectors to prevent resolution deterioration caused by large capacitance. Fabrication of a Li drifted large area detector requires complicated techniques, but UHP p-Si fascilitates simple and easy fabrication process. A Surface Barrier Detector of 20 cm^2 electrode area was fabricated with 50 KΩ·cm p-type Si Wafer of 64 mmϕ (13).

The spectrum of Alpha and Beta-rays injected together to the detector is shown in Fig. 9. The energy resolution in FWHM is 36 keV for Alpha particles, 33 keV for 1 MeV electrons and 30 keV for test pulses. The leakage current at 400 V bias was 6 uA and remained almost constant against increasing bias. The depletion thickness was approximately 2 mm according to the capacity measurement. The electrode fringe was protected by epoxy resin and encapsulated in a metal case. Since only the Al electrode of the wafer was exposed, the detector was insensitive to light, and the strong adhesivity of Al to Si facilitates the routine use of a large detector in our institute as a surface contamination monitor and Alpha and Beta-rays detector for environmental samples of very low activity.

5. DISCUSSION

Among the S.B.D.'s of UHP p-Si, partially depleted detectors of low leakage current and high breakdown voltage characteristics could be successfully fabricated for both room and liq. N_2 temperature use. By selecting appropriate surface chemical treatment, the leakage current of a 1 mm thick detector can be reduced down to 0.2 uA/cm^2 at a 100V bias. Even when the detectors are cooled to liq. N_2 temperature, the breakdown voltage characteristics of these partially depleted detectors do not deteriorate, as they often do in n-type Si S.B.D.'s.

The main reason for this difference may be that the inversion layer is more easily and naturally formed on a p-Si surface than on n-Si. As for totally depleted detectors, thin detectors show good characteristics with only a few exceptions: -- low leakage current, high breakdown voltage, no charge injection at the surface electrode; while for thick detectors, there remains the problem of a sharp leakage current increase at liq. N_2 temperature (14). The back contact (Au electrode side) is the most probable cause for the marked response. Surface chemical treatment plays a key role in S.B.D. fabrication, and the treatment markedly influences the detector characteristics. Thus, experimental investigation with respect to this problem has to be continued. Adding a guard ring or a groove to the electrode structure has not been tried, but it is quite possible to improve the detector characteristics by these modifications.

As an application of p-type Si crystal to radiation detectors, a P-N

diffused junction detector was fabricated by a planar method with moderately high resisitive p-type Si (15). The planar method has not yet been tested with UHP p-Si. As another electrode preparation technique, the ion implantation method also seems promising to produce a thin entrance window, but this technique has not yet been tested with UHP p-Si. Surface passivation too, has not yet been examined closely and is leading to farther experimental studies. According to our knowledge, thermal treatment is needed following the implantation,then special care has to be taken to prevent the contamination of UHP p-Si from ambient impurities.
Apart from the use in radiation detectors, UHP p-Si may be used as a starting material for nentron transmutation doping (N.T.D.). The ideal starting materal must have a very low concentration and an excellent uniform didtribution of acceptor impurities; UHP p-Si fulfills these condition satisfactorily. Surface Barrier Detectors of good performance characteristics have been fabricated using n-type Si produced by the N.T.D. method with this material (16).

Considering weak points for detector material, it must be mentioned that (1) hole trapping effect is observed in some UHP p-Si's, and

(2) accurate measurement of extremely high resisitivity is difficult to be obtained. Hole trapping effect is observed in samples of the early developmental stage, and its cause is not clarified even now. Measurement of UHP p-Si resisitivity of more than 100KΩ·cm at room temperature is an extremely difficult task. Fabricating a large amount of detectors with UHP p-Si, it was often noticed that even if their resisitivity and the applied bias were alike, some depleted quite deeply while others did not. To faricate a large quantity of thick detectors, e.g., above 10 mmt, the resisitivity must undoubtedly be measured as accurately as possible. But this value alone is not sufficient to predict the depletion thickness of the detectors. Other factors, some of which are still unknown, have yet to be investigated. The resisitivity may be estimated through detector depletion thickness using a known bias, but there remains the problem of reproducibility for S.B.D.'S of the present surface chemical treatment. Fundamental experimental work and the development of a theory of depletion layer formation on UHP p-Si must be carried out to understand and improve this UHP p-Si.

6.CONCLUSION

UHP p-Si crystal can be utilized to fabricate S.B.D.'s of various uses, and its fabrication technique is simple and easy. Partially depleted detectors and totally depleted detectors of less than 4 mmt for room temperature measurement can be fabricated without difficulty. Some problems however remain unsolved. These are : surface chemical treatment for cooled detectors and crystal properties for thick totally depleted detectors. But these problems do not seem to be too difficult and may be solved in the near future. The utilization of UHP p-Si in radiation detectors is only in its begining. Only S.B.D.'s have been developed and studied. Even for S.B.D.'s the ion implantation method and many other techniques have not yet been tested. The production of UHP p-Si crystal is still at a research level and its difficult acquisition is still a problem in the further process of experimental investigations with this material. To increase the utilizations with this material, an extensive range of experiments will be necessary until the definite characterization of this interesting crystal may be obtained.

REFERENCES

(1) A. Yusa, Y. Yatsurugi and T. Takaishi; J. Electrochem. Soc.122, 1700 (1975)
(2) D. Itoh, I. Namba and Y. Yatsurugi; in this symposium
(3) M. Tajima; Appl. Phys. Lett. 32 (11), 1 ,June (1978)
(4)P.J. Mathew, N.G.Chapman and G.E. Coote; Nucl. Instr. and Methods, 49, 245 (1967)
(5) Yu. S. Maksimov, Yu. F. Rodinov,and Yu. N. Yavlinski; Soviet Phys. Semiconductors, 1,No. 7, 867 (1968)
(6) R. Chaudhry and R.V. Srikantiah; B.A.R.C.-527 (1971)
(7) A.V. Protsenko,V.N.Sinitsyn, N.V.Panasenko, and V.M.Korol;Soviet Phys,-Semiconductors 3, No. 9, 1118 (1970)
(8) E. Elad, C.N. Inskeep, R.A. Sareen, and P. Nestor; IEEE NS-20, No.1, 534, (1973)
(9) V.F. Kushniruk, R.A.Nikitina, and Yu. P. Kharitonov; Soviet Phys.-Semiconductors 7, No.7, 933 (1974)
(10) V.V. Avdeichikov, E.A. Ganza, and V.P.Prikhodtseva; Nucl. Instr. and Methods 133, 579 (1976)
(11) Y. Takami, F.Shiraishi, and M. Hosoe; to be published in IEEE NS-30, No. 1
(12) F.Shiraishi, M. Hosoe, Y.Takami and Y. Ohsawa ; IEEE NS-29,No.1, 775(1982)
(13) Y. Takami, M. Hosoe, and F. Shiraishi; to be published.
(14) F. Shiraishi and Y. Takami; Nucle. Instr. and Methods 196, 137 (1982)
(15) M.Yabe, N. Sato, H. Kamijo, T. Takechi, and F. Shiraishi; Nucl. Instr. and Methods 193, 63 (1982)
(16) C.Kim, H.W.Kraner, D. Itoh, K. Husimi, S. Ohkawa, and F. Shiraishi; Nucl. Instr. and Methods, 196, 143 (1982)

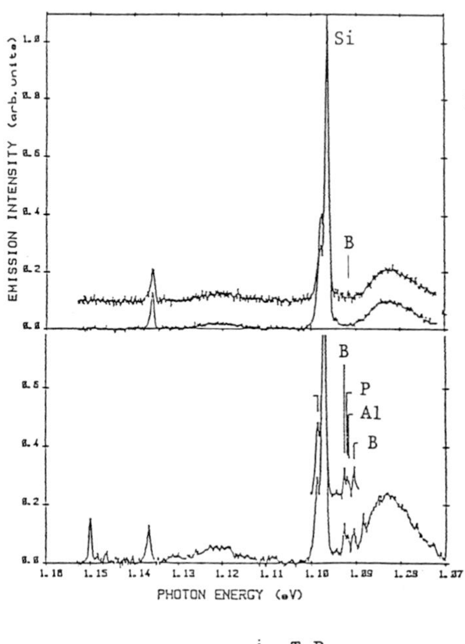

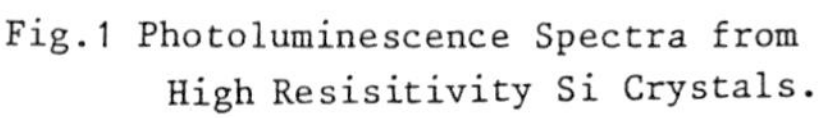
Fig.1 Photoluminescence Spectra from High Resisitivity Si Crystals.

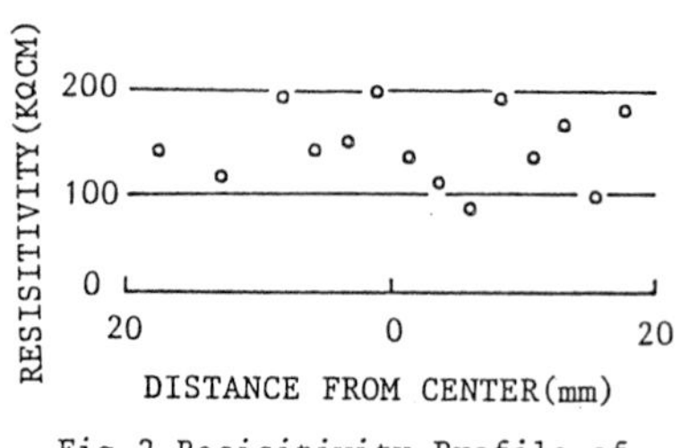

Fig.2 Resisitivity Profile of UHP p-Si.

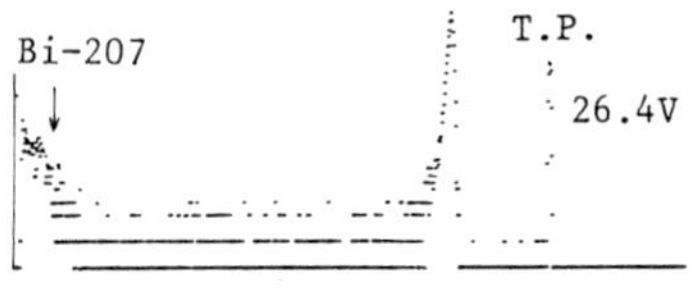

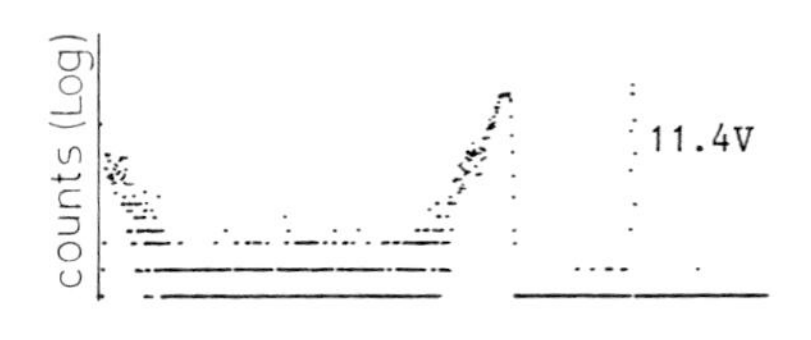

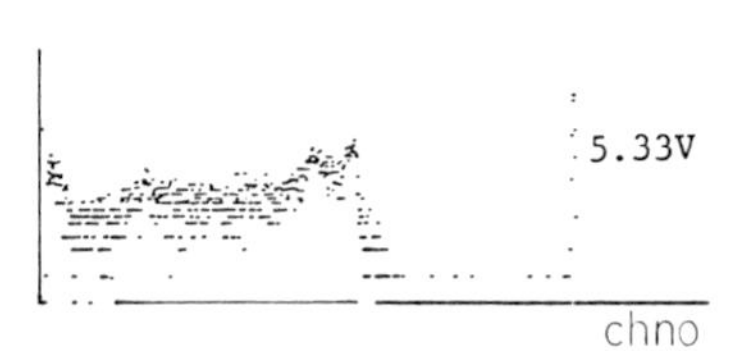

Fig.3 Pulse Height Saturation Behaviour of a p-Si Surface Barrier Detector(1 mmt) at Different Bias and 23°C. Alpha Particles were Injected to the Al-face.

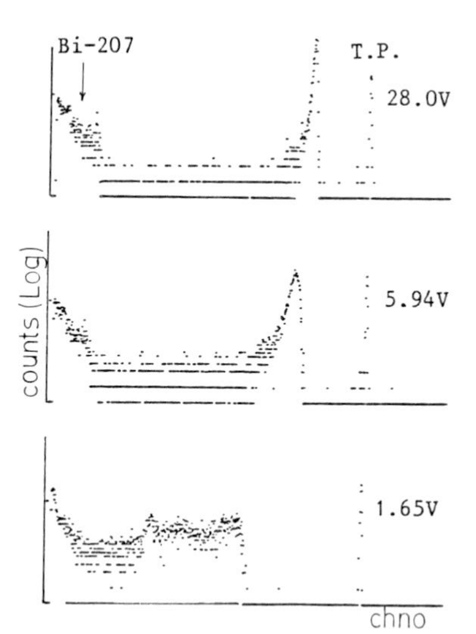

Fig.4 Pulse Height Saturation Behaviour of a p-Si S.B.D.(1 mmt) at Different Bias and at 23°C. Alpha Particles were injected to Au- face.

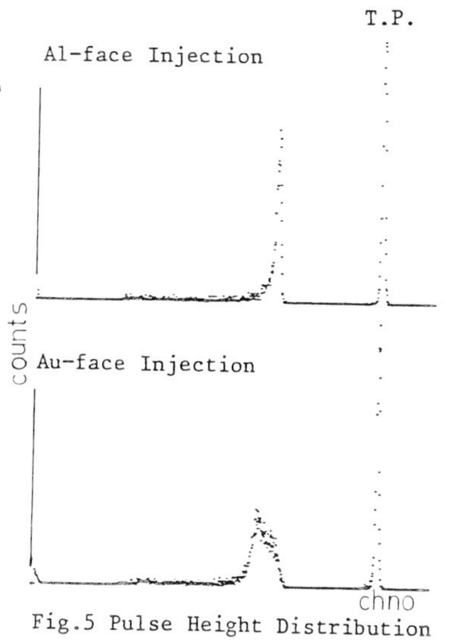

Fig.5 Pulse Height Distribution of Alpha Particles Obtained by 12.3 mmt S.B.D. at 2.8 KV Bias and at 0°C.

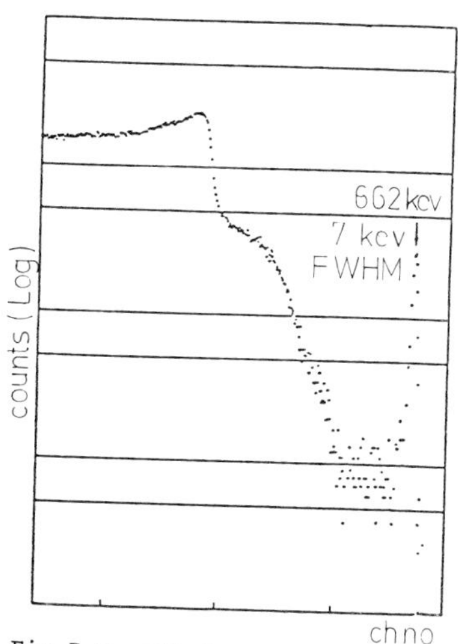

Fig.7 Cs-137 Gamma-ray Spectrum Obtained by 15 mmt S.B.D. at 77K.

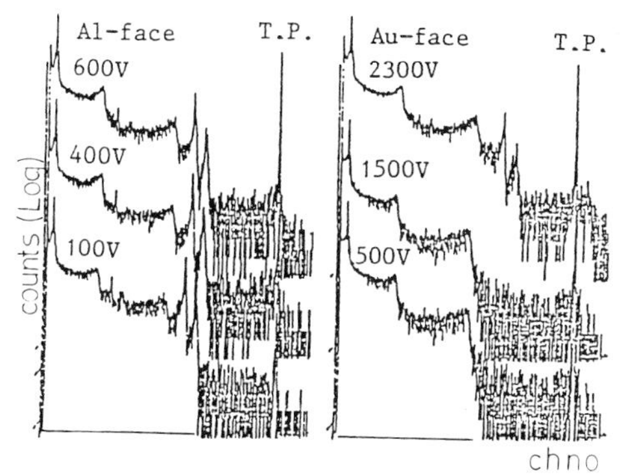

Fig.6 Bi-207 Internal Conversion Electron Spectra Obtained by 15 mmt S.B.D. at 77K. To Al-face Injection, Line Spectra can be Observed at Rather Low Bias Region (but 90% of the saturated pulse height even at 2.3 KV); While to Au-face Injection, the Peaks of Good Resolution are Observed at Higher Bias of 2.3 KV (100% of the saturated pulse height).

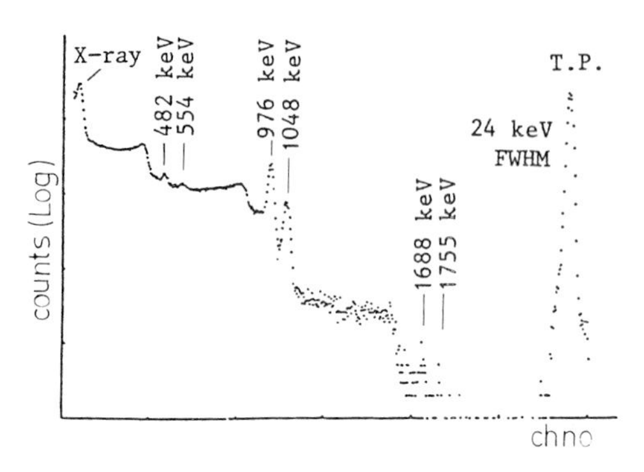

Fig.8

Bi-207 Internal Conversion Electron Spectrum Obtained by a Partially Depleted Detector of 1 cm^2 Electrode Area at 300V Bias and at 25°C. The Energy Resolution is 30 keV FWHM at 976 keV Peak.

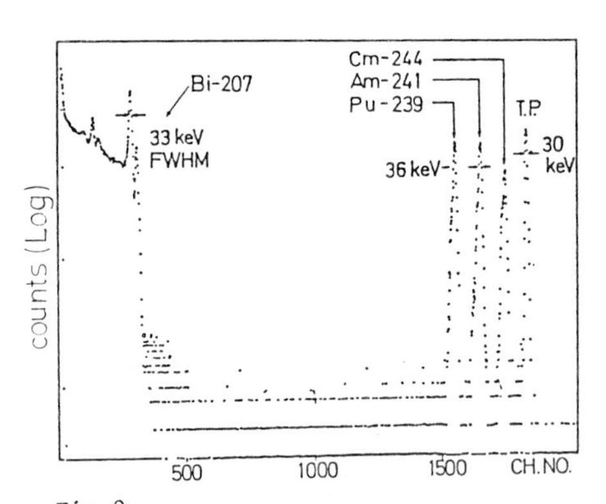

Fig.9

Spectrum of Bi-207 Electrons and Alpha Particles Measured by a Large Area S.B.D. (20 cm^2) at 25°C.

HIGH-VOLTAGE OPERATION OF SURFACE BARRIER SILICON DETECTORS WITH A SIDE GROOVE

S. OHKAWA AND K. HUSIMI
Institute for Nuclear Study, University of Tokyo, Tanashi, Tokyo, Japan
C. KIM AND Y. KIM
Department of Electronics Engineering, Korea University, Kodaira, Tokyo, Japan
D. ITOH
Komatsu Electronic Metals Co., Hiratsuka, Kanagawa, Japan

ABSTRACT

A remarkable characteristic of the surface barrier detector with a side groove cooled down to the nitrogen temperature has been observed. That is an abrupt decrease of the output noise at a definite voltage in increasing the bias voltage slowly. This is caused by a carrier injection from the side surface of the groove near the neck narrowed by the side groove. This excess noise disappears after the depletion layer goes through the neck of the side groove. This is confirmed by the fact that the capacitance of the detector decreases abruptly with the decrease of the noise at the same bias voltage.

This detector is capable to be operated at a voltage higher than this voltage with a low noise. The maximum bias voltage applied to the detector is 3000 V.

INTRODUCTION

The T type structure with a cylindrical groove has widely been used for producing various kinds of thick silicon detectors. This conventional groove is effective for increasing the bias voltage of a thick detector, because the field surrounding the electrode is pinched off by the neck of the groove [1,2,3,4,5].

It has pointed out however, that in lithium compensated silicon detectors surface channels on the wall of the groove produce field distortions in the bulk and this causes the charge due to events produced near the wall of the groove to be partially collected in the surface layers [6,7]. This "disturbed region" decreases the effective window area of the detector considerably [8].

In order to reduce this "disturbed region" and to apply a high bias voltage to the surface barrier detector made by using a high purity silicon crystal, a groove is prepared perpendicular to the side wall. This side groove is expected to be effective for increasing the length of the path of the surface channels and for decreasing the field surrounding the surface barrier electrode, which are the same functions of the conventional cylindrical groove.

FABRICATION OF THE DETECTOR

The construction of the surface barrier silicon detector with the side groove is shown in figure 1. The side groove is cut by using the tool shown in figure 2. The silicon wafer is fixed on the disk by a wax. The disk rotates at a speed of 1000 rpm driven by a moter. The knife of the cutter is

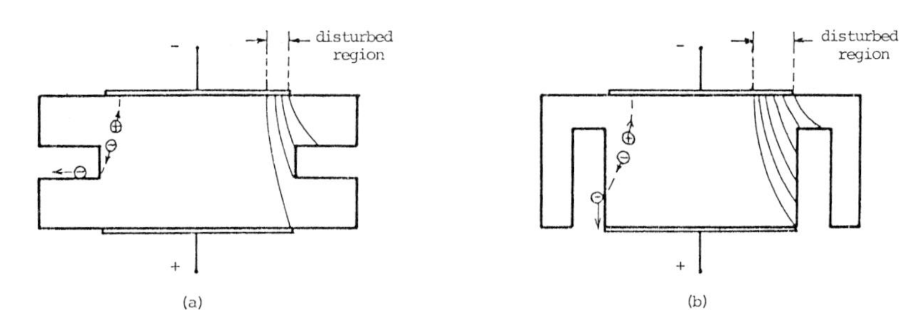

Fig. 1 Constructions, profiles of electric fields and flows of ionized carriers in surface channels
a): detector with the side groove
b): detector with the cylindrical groove

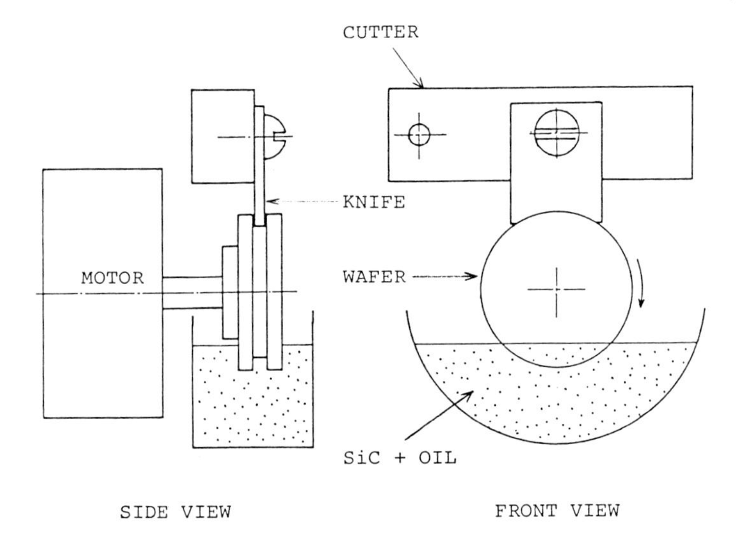

Fig. 2 Grinding tool for producing the side groove on a silicon wafer

a stainless steel plate, the thickness of which determines the width of the groove. The grinding material is a carborundum powder of mesh #1000 suspended by a machine oil. The knife grinds the silicon wafer by the own weight.

The silicon wafer has a diameter of 18.5 mm and a thickness of 5.1 mm. The depth and the width of the groove are 3.8 mm and 2.4 mm respectively. The resistivity of the wafer is estimated to be 40 kΩ-cm (N-type) from the capacitance-voltage relationship. The surface barrier electrode has a area of 65.5 mm^2 and is produced by evaporation of gold in vacuum. The ohmic electrode with a area of 95 mm^2 is also produced by evaporation of aluminium.

CHARACTERISTICS OF THE DETECTOR

The current voltage characteristics of this detector measured at liquid nitrogen temperature is shown in figure 3. The current increase as a function of applied voltage becomes slow in the region from 700 V to 1500 V. In this region, the bottom of the depletion layer is supposed to be pinched within the neck of the groove. The current increase becomes steep again above the voltage higher than 2000 V. The maximum bias voltage capable to be applied is 3000 V.

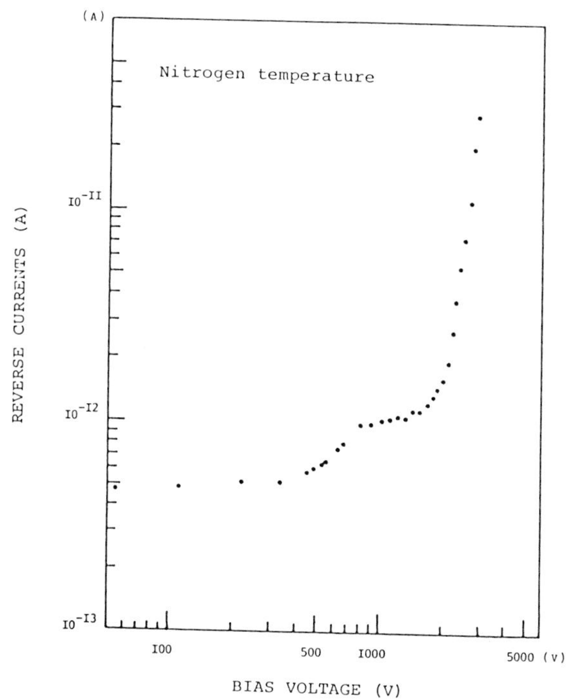

Fig. 3 Reverse current characteristics of the detector with the side groove

The characteristics of noise of the detector is obtained as a function of the applied voltage by measuring the output signal of the main amplifier using a voltmeter. Figure 4 shows the result, in which the voltage is calibrated by the pulser resolution. As seen from the figure, the excess noise is observed in the region of the bias voltage from 300 V to 500 V. The noise decreases abruptly at the voltage of 500 V.

The capacitance of the detector is measured as a function of the bias voltage by means of the pulser method as shown in figure 5. The capacitance also decreases steeply at the same voltage as shown in the figure. The bias voltage dependence of the capacitance is smaller than that expected from $V^{0.5}$ law in the region under the bias voltage of 500 V and it becomes larger than that in the region above 500 V.

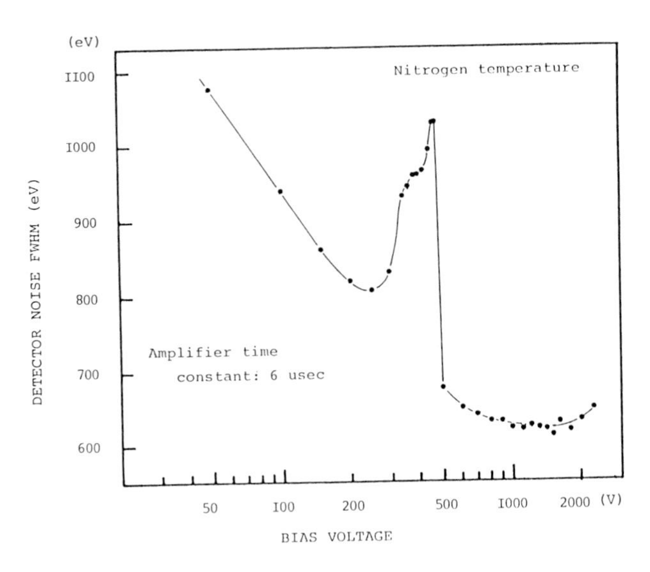

Fig. 4 Noise of the detector with the side groove as a function of the bias voltage

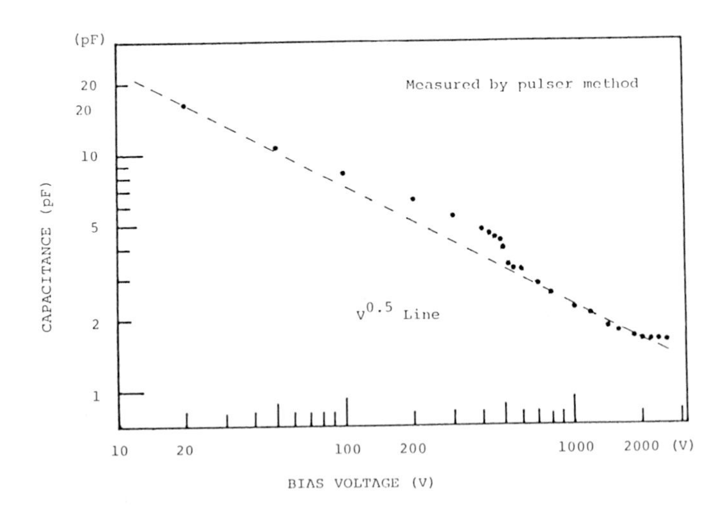

Fig. 5 Capacitance of the detector with the side groove as a function of the bias voltage

Figure 6 shows the spectrum of X-ray from ^{241}Am collimated by a slit with a hole of 5 mm obtained by this detector, the bias voltage of which is 2200 V. The energy resolution of Np Lα1 peak (13.95 keV) is 653 eV (FWHM).

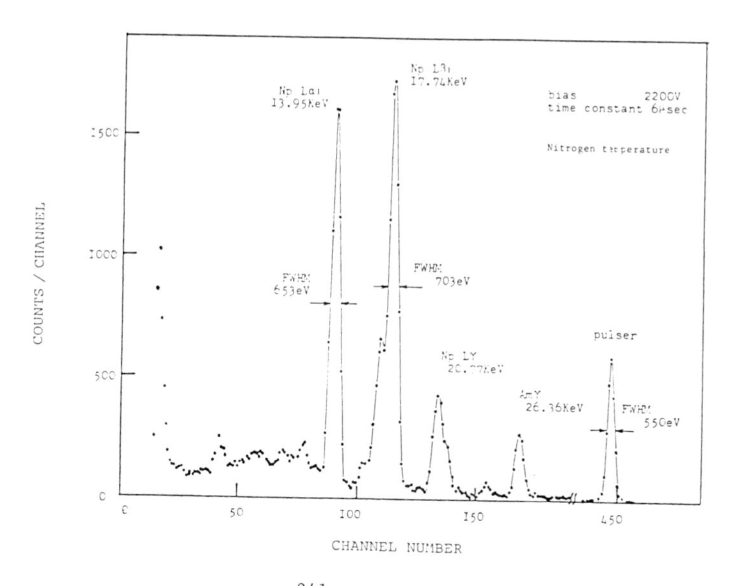

Fig. 6 Spectrum of X rays from ^{241}Am obtained by the detector with the side groove

CONCLUSIONS

It is shown from the capacitance measurement that the bottom of the depletion layer arrives in the entrance of the neck of the side groove at the voltage of 500 V. At the bias voltage slightly lower than 500 V, the bottom of the depletion layer approaches the side wall of the groove and even partially touches the wall. After the bottom of the depletion layer touches the wall completely and the bias voltage exceeds 500 V slightly, the bottom of the depletion layer extends into the neck of the side groove. From the catastrophical decrease of the excess noise, it will be supposed that a redistribution of the field in the depletion layer would occur at this critical voltage. However, this is necessary to be confirmed later.

There are two shelves separated by the side groove in this type of detectors. The shelf of the side of the surface barrier electrode is considered to have the same pinching off function of the electric field surrounding the surface barrier electrode as the fringe of the conventional cylindrical groove. On the other hand, the shelf of the side of the ohmic electrode is effective for increasing the length of the path of the surface channels from the ohmic electrode to the neck of the groove. This will be favourable for decreasing the "disturbed region" caused by the charge collection in the surface channels compared with the detector having the conventional cylindrical groove.

REFERENCES

1. J. Llacer, IEEE Trans. Nucl. Sci., NS-11, No. 3, 221 (1964)

2. F. S. Goulding, Nucl. Instr. and Meth. 43, 1 (1966)

3. J. Llacer, IEEE Trans. Nucl. Sci., NS-13, No. 1, 93 (1966)

4. R. J. Fox and C. J. Borkowski, IEEE Trans. Nucl. Sci., NS-9, No. 3, 213 (1962)

5. R. A. Ristinen, D. A. Lind and J. L. Homan, Nucl. Instr. and Meth., 56, 55 (1967)

6. H. L. Malm and R. J. Dinger, IEEE Trans. Nucl. Sci., NS-23, No. 1, 76 (1976)

7. F. S. Goulding, Nucl. Instr. and Meth., 142, 213 (1977)

8. S. Hessel, IEEE Trans. Nucl.Sci., NS-29, No. 1, 751 (1982)

CHARACTERIZATION OF SINGLE CRYSTAL MERCURIC IODIDE (HgI_2) USING THICK DETECTOR STRUCTURES*

A. BEYERLE, K. HULL, J. MARKAKIS, W. SCHNEPPLE, AND
L. VAN DEN BERG
EG&G, Energy Measurements Group, Santa Barbara Operations,
130 Robin Hill Road, Goleta, California 93117

ABSTRACT

Single crystal sections of HgI_2 (about 1 cm thick) have been evaluated for charge carrier transport properties. Using these thick detector structures, surface effects produced during fabrication are reduced, enhancing the bulk property characteristics. The standard time-of-flight method was used to determine electron and hole mobilities. Lifetime measurements for electrons and holes were made by direct observation of the carrier decay where crystal transit times were long compared to lifetimes. Nonlinear charge carrier velocities have been observed during mobility measurements. These nonlinearities impair a partial charge collection technique [1,2,3] for spectra generation taken with thick detectors while having a lesser effect on the standard full charge collection approach. Partial charge collection methods have produced greater peak efficiencies than full charge collection, and investigations of crystal properties are being used to enhance this method.

INTRODUCTION

Until recently, mercuric iodide (HgI_2) has been limited to relatively thin (<1 mm) detector structures [4]. Thick sections of HgI_2 are now available with sufficient quality to fabricate 1-cm-thick detector structures having reasonable charge transport efficiency. Thin detectors of HgI_2 serve as fine x-ray spectrometers but their efficiency is a serious limitation for gamma-ray work. One-cm-thick detectors offer fairly good efficiency at moderately high gamma-ray energies (~30% at 662 keV [5]) which makes HgI_2 a reasonable candidate for such a detector if gamma-ray energy information can be extracted from the crystal. For this reason, work to develop the potential of HgI_2 as a gamma-ray spectrometer is underway, and studies of the material properties of HgI_2 is an essential part of this effort.

---

*This work was performed under the auspices of the U.S. Department of Energy under Contract No. DE-AC08-76NV01183.

DETECTOR FABRICATION

Large crystals of HgI_2 weighing up to 400 g are being grown at EG&G from the vapor phase with techniques described elsewhere [6]. The crystals are sawn into blocks having active areas from 2 to 10 cm^2 and thicknesses from 0.4 to 1.2 cm. Detectors are fabricated from these single crystal blocks by applying electrodes to the large area 'C' faces, either by painting on a carbon suspension solution (Aquadag) or by vacuum deposition of palladium. Figure 1 shows a typical 1-cm-thick detector of the type used for the measurements presented here. Further details on the fabrication process have been presented by Whited and Schieber [7]. The more complicated process of applying evaporated contacts is necessary in those cases where the detector is to be irradiated with alpha particles for certain charge transport property measurements.

Figure 1. Typical mercuric iodide detector of the type used for these measurements.

APPARATUS AND PROCEDURE

In the effort to improve the quality of mercuric iodide thick-detector spectra, the detector output pulses have been studied with the apparatus shown schematically in Figure 2. For most measurements the results were obtained in this manner by just investigating the charge-sensitive preamplifier output. Detectors are placed in light-tight electrically shielded fixtures and allowed to remain under bias for several days before measurements begin. A series of measurements on a detector can take several months because the detector must be allowed to stabilize any time the bias is changed. Detector output waveforms are recorded on a storage oscilloscope or a Tektronix R7912 transient digitizer system. The spectrum shown was obtained by using a standard nuclear gaussian shaping amplifier on the preamplifier pulses with shaping constants short ($\sim$0.5 μsec) compared to the charge collection times.

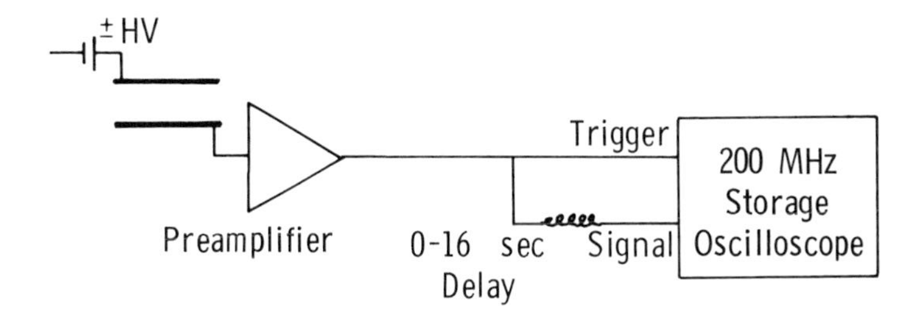

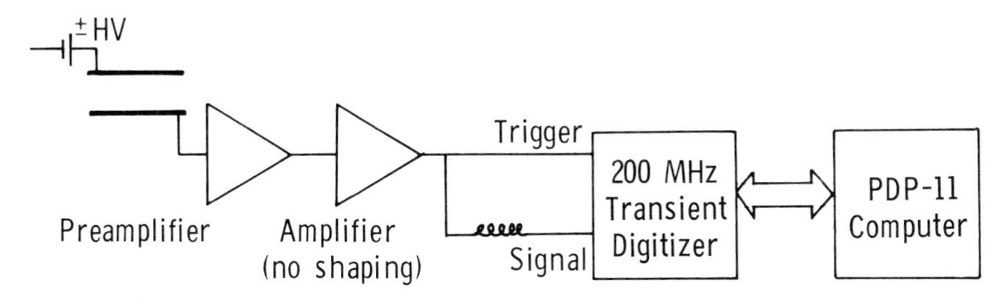

Figure 2. Schematic of two electronics configurations to study detector output pulses.

MOBILITY MEASUREMENTS

The velocity ($v = \mu E$) of charge carriers in a crystal material is determined by the electric field inside the crystal ($E \sim V/d$) and the carrier mobility (μ). By simply measuring the time (t_r) it takes for the carriers to transit the detector structure the mobility ($\mu = d^2/Vt_r$) of the carriers is found.

The usual method of measuring the transit time by this technique is to produce electron hole pairs near the detector surface with energetic alpha particles through a thin contact. We have used a ^{244}Cm alpha source ($E_\alpha \sim 5.8$ MeV, $E\gamma = 43$ keV) with thin evaporated palladium contacts. With detectors this thick, another technique is to use low-energy gamma radiation. The ^{241}Am gamma-rays ($E\gamma = 60$ keV) penetrate less than a millimeter into the detector, so that all of the carriers are produced near the surface. The gamma radiation can be used with the painted carbon contacts, which the alpha particles will not penetrate.

Two electron-induced pulses are shown in Figures 3a and b. The majority of the electrons are tranversing the crystal at 5000-V bias (Figure 3b) such that the pulse is quite a straight line until the electrons are collected. For electrons the transit times were usually found to vary between 1 and 3 μsec, which implies mobilities of about 100 cm^2/V-sec. Values of electron mobilities on the order of 100 cm^2/V-sec are in agreement with the best previously published values [7]. Two detectors cut from a single crystal showed exceptionally high electron mobilities on the order of 300 cm^2/V-sec. While this is of course a desirable property to produce in detectors, it is hard to draw conclusions about growth factors from one crystal.

Hole mobilities are slightly more difficult to measure because hole collection across this distance is incomplete. We have used a thinner crystal of 4.2 mm rather than 10 mm. At high biases with this thinner crystal we can get a significant portion of the holes across the crystal such that by observing t_r directly the hole mobility (μ_h) can be reasonably estimated.

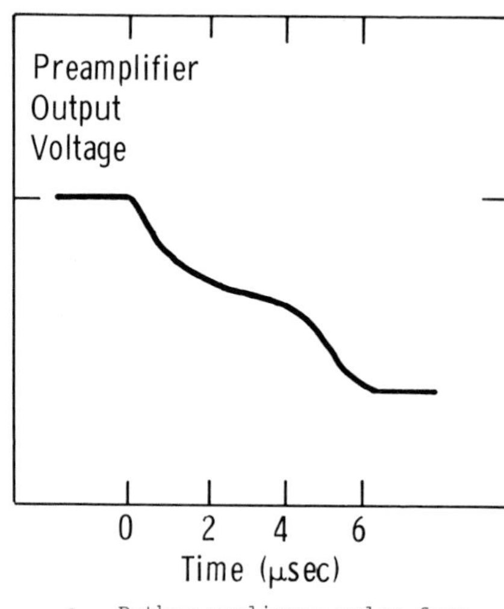

a. Rather nonlinear pulse from detector with 2 kV bias.

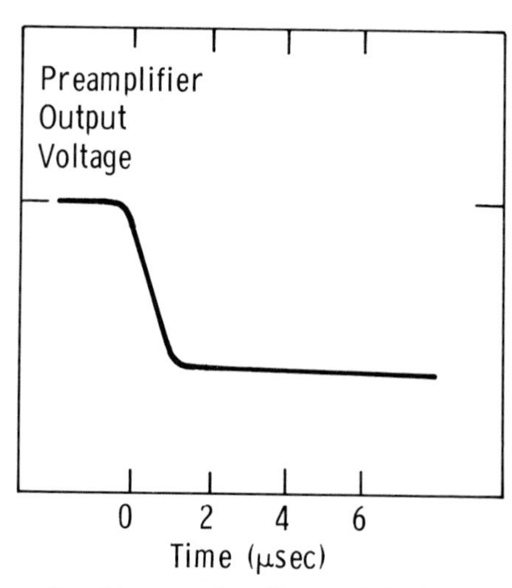

b. Linear pulse from a detector with 5 kV bias.

Figure 3. Two electron-induced pulses from thick HgI_2 detectors

It is known that the bias is high enough to be measuring mobility rather than lifetime when the μ_h value no longer changes with increases in bias. In our measurements taken on the 0.42-cm detector operating at 4-5 kV, we have found $\mu_h \sim 6$ cm^2/V-sec, which is in agreement with previously measured values [7].

CARRIER LIFETIME

For low enough electric fields inside the crystal, the charge carriers will no longer traverse the crystal in the carriers' lifetime (τ_e, τ_h). For this case the shape of the output pulse can be used to measure τ_e and τ_h, rather than t_r as described above.

For holes in a 1-cm-thick crystal there is no difficulty providing conditions under $\tau_h < t_r$, because the mobility is low and the lifetimes short. Pulse rise time measurements have been taken for many alpha-particle-induced pulses near the positive electrode, so that the holes must induce most of the charge. At the lowest biases, the charge collection time depends on the lifetime of the holes, and not on bias. In this manner $\tau_{1/2}$ is found to be approximately 12 μsec, and $\tau = \tau_{1/2}/\ln(2)$ which implies $\tau \sim 15$ μsec. The result of the hole measurements yields a $\mu\tau_h$ product of about 9×10^{-5} cm^2/V, which is slightly better than values around 3×10^{-5} cm^2/V previously reported. Even this slightly higher value yields only a hole trapping length $\tau_h \sim 3$ mm, which is the reason we must use a thinner crystal for the hole mobility measurements mentioned above.

Electron lifetime measurements are difficult by this method because the lifetime is so long. We have been able to establish that the electrons are still active after 200 μsec. With voltages low enough to cause $t_r \sim 200$ μsec, the rate of charge induction on the electrodes is small. Preamplifier noise has limited our ability to measure charge collection spread out over a longer period of time. Previous measurements of $\mu_e \sim 100$ cm^2/V-sec and $(\mu\tau)_e \sim 5 \times 10^{-4}$ cm^2/V [7,8,9] yield electron lifetimes $\tau_e = 5$ μsec. These $\mu\tau$ measurements depend on charge collection efficiency $\eta = Q/Q_o$ measurements on much thinner detector structures.

Although present results are quite different than previous, the present work relies on a different technique to make a direct measurement of τ. Another direct measurement of electron-induced charge decay was conducted at University of Southern California by Szymczyk, et al, [2] who found "at least several hundred μsec" conduction lifetime in HgI_2.

This result indicates that a surface effect may be playing a role in charge collection. A large variation in $\mu\tau$ values that depends systematically on detector thickness has been found by Levi, Schieber, and Burshtein [10] in their surface effects studies on HgI_2 crystals ranging from 0.105 to 1.43 mm in thickness. The thicker the detector the higher the lifetime that is observed. They conclude that "electron surface recombination ... dominates over bulk trapping" and that "some $\mu\tau$ estimates given in the literature are quite doubtful." The present measurement of >200 μsec electron lifetime and the USC measurement [2] of "at least several hundred μsec" tend to support this argument.

Table I summarizes the μ and τ measurements from present work, and the τ measurement is displayed as a $\mu\tau$ product for comparison with other work. The present effort yields values slightly higher than previous work probably due to our use of thicker crystals.

Table I. Comparison of charge carrier property measurements.

	μ_e cm²/V-sec	$\mu_e\tau_e$ cm²/V	μ_h cm²/V-sec	$\mu_h\tau_h$ cm²/V
Present Work	100	10^{-2}	6	9.0×10^{-5}
"Best Value"[7]	100	5×10^{-4}	4	3.0×10^{-5}
Schieber[8]		4×10^{-5}		4.0×10^{-6}
Whited[11]		2×10^{-4}		1.5×10^{-5}
USC 1978[9]		$\left\{\begin{matrix}5 \times 10^{-5}\\6 \times 10^{-4}\end{matrix}\right\}$		$\left\{\begin{matrix}8.0 \times 10^{-7}\\7.0 \times 10^{-6}\end{matrix}\right\}$
Minder[12]	100		4	

CRYSTAL UNIFORMITY

Mobility and lifetime measurements by the means described above assume that the material is uniform or that the quantity measured is an average over the entire volume of the crystal. Additional information can be obtained from the output of thick detectors on the depth uniformity of the crystals. The instantaneous rate of charge induction on the electrodes is $dQ/dt = n_e\, q_e\, \mu_e\, E/d + n_h\, q_h\, \mu_h\, E/d$. In two electron-induced output pulses from HgI_2 detectors shown in Figure 3, the first pulse (3a) is highly nonlinear and the other has a uniform rate of charge collection. The nonlinear outputs are associated with detectors that have not had a conditioning period [3] of time to rest under bias, radiation conditioning, or light treatment. This suggests there is a time period of several days before the detector reaches equilibrium, and the detectors must be irradiated to maintain the proper state of equilibrium. Nonlinear detector output pulses are also associated with lower biases. Much more success is obtained in achieving linearity with biases above 4000 V/cm.

SHORT CHARGE COLLECTION TIME SPECTROMETRY

The motivation behind measurements of charge transport properties in terms of the present work is the development of gamma-ray spectrometers with HgI_2. Since $dQ/dt = n_e\, q_e\, \mu_e\, E/d + n_h\, q_h\, \mu_h\, E/d$, the rate of charge collection depends only on the velocities $v = \mu E$ of the carriers and the electric field, as the electron charge and detector thickness do not change. This rate continues so long as the charge carriers are free to drift in the crystal. If the charge induced on the electrodes is sampled for a fixed short period of time Δt after an event, the charge collected $Q = n_e q_e \Delta x_e/d + n_h q_h \Delta x_h/d$, where $\Delta x_{e,h} = v\Delta t = \mu E \Delta t$, is the distance traveled by the electrons and holes in Δt. If we have μ and E sufficiently uniform over the entire volume of a detector, the rate of charge collection is dependent only on $n_{e,h}$. The definition of sufficiently uniform crystal parameters and the means to obtain these conditions are the major impetus for this work. Since the hole mobility in mercuric iodide is about 1/25 the electron mobility, the hole-induced charge contribution is small. Figure 4 shows schematically a set of gamma-ray-induced events in a detector and the corresponding output pulses, with the finite lifetime of the carriers ignored. The energy information about the gamma-ray ($\propto n_{e,h}$) is best obtained very soon after the event (short charge collection) or by the more usual long charge collection time. For 1-cm-thick HgI_2 detectors the long charge collection time is longer than τ_h, which has been the limitation on thickness of spectrometers.

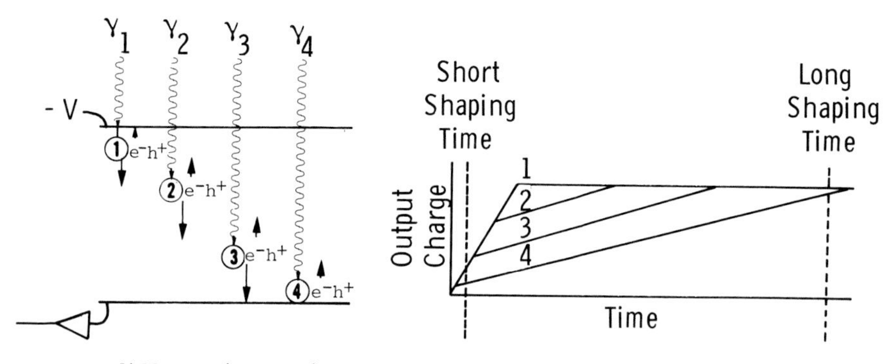

a. Four different interaction positions of a gamma-ray.

b. Corresponding preamplifier output pulses.

Figure 4. Effects of interaction position on preamplifier output.

The short charge collection time method has been used with some success [1]. With short charge collection, there is a compromise between the longer collection times for best signal-to-noise and shorter collection to satisfy the condition that the electrons remain free to drift and are not collected. The short charge collection has been implemented by simply using 0.5-μsec differentiation and integration constants on a standard nuclear amplifier. The 0.5-μsec collection time causes a 'dead layer' of about 20% of the crystal for which the energy registration is incorrect (corresponding to Event 4 in Figure 4). A spectrum obtained with a 1-cm-thick HgI_2 detector and a ^{137}Cs source is shown in Figure 5. Resolution at full width half maximum (FWHM) is about 15% for the 662-keV gamma-ray. Total efficiencies are within 15% of the expected value of 25%. Peak efficiencies are about 30% of the total efficiency, which is above the photoelectric value for this energy.

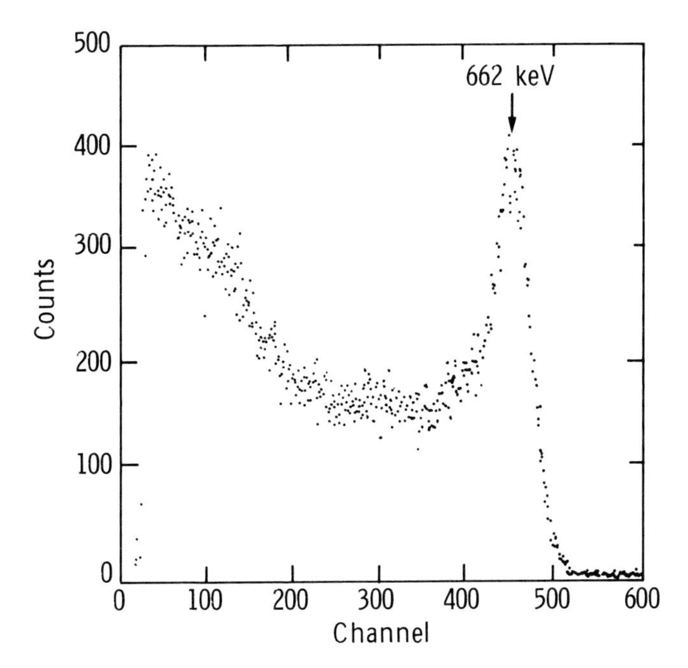

Figure 5. ^{137}Cs spectrum from a 1-cm-thick HgI_2 spectrometer with 0.5-μsec shaping time.

CONCLUSION

The charge carrier transport property measurements for the most part have tended to confirm previous measurements of μ. The τ measurements yield values higher than those derived from previous $\mu\tau$ measurements, especially for the case of τ_e. This difference may be explained by the surface effects which have hampered previous $\mu\tau$ values inferred from η measurements, since the bulk properties are much more important in these much thicker crystals. The present measurements as well as the spectra which have been obtained [1] show an ability to collect electrons efficiently through 1-cm blocks of HgI_2. As thicker detectors of HgI_2 become available, the limit on τ_e may be lengthened.

The present work on thick mercuric iodide detectors has resulted in gamma-ray spectrometers of much greater efficiency than has been obtained in the past. The application of the short charge collection time technique has enabled the extraction of gamma-ray energy information from more events than the more usual long shaping time techniques. The low mobilities, great difference in electron and hold mobilities, low leakage currents, and long electron lifetimes have made the application of this technique practically possible.

ACKNOWLEDGEMENTS

The authors wish to thank Carole Ortale and the EG&G staff of crystal growers and detector fabricators for their help and advice concerning fabrication of all detectors used in this work, and M. Szymczyk of USC for many useful discussions concerning the technical aspects of these measurements.

REFERENCES

1. A. Beyerle, K. Hull, J. Markakis, and B. Lopez, Nucl. Instr. and Method., to be published in special issue on Jerusalem Workshop on HgI_2.

2. W.M. Szymczyk, A.J. Dabrowski, J.S. Iwanczyk, J.H. Kusmiss, G.C. Huth, K. Hull, A. Beyerle, and J. Markakis, Nucl. Instr. and Meth., to be published in special issue on Jerusalem Workshop on HgI_2.

3. K. Hull, A. Beyerle, B. Lopez, and J. Markakis, IEEE Transactions, to be published.

4. C. Ortale, L. Padgett, and W. Schnepple, Nucl. Instr. and Meth., to be published in special issue on Jerusalem Workshop on HgI_2.

5. H.L. Malm, IEEE Trans. Nucl. Sci., NS-22, 182 (1975).

6. L. van den Berg and W.F. Schnepple, Proceedings of the Materials Research Society 1982 Annual Meeting, Boston, Massachusetts, this issue.

7. R.C. Whited and M. Schieber, Nucl. Instr. and Meth., 162, 113 (1979).

8. M. Schieber, I. Beinglass, G. Dishon, A. Holzer, and G. Yaron, IEEE Trans. Nucl. Sci., NS-25, 644 (1978).

9. A.J. Dabrowski and G. Huth, IEEE Trans. Nucl. Sci., NS-25, 205 (1978).

10. A. Levi, M.M. Schieber, and Z. Burshstein, Nucl. Instr. and Meth., to be published in special issue on Jerusalem Workshop on HgI_2.

11. R.C. Whited and L. van den Berg, IEEE Trans. Nucl. Sci., NS-24, 165 (1977).

12. R. Minder, G. Majni, C. Canali, G. Ottaviani, R. Stuck, J.P. Ponpon, C. Schwab, and P. Siffert, J. Appl. Phys., 45, 5074 (1974).

STATE-OF-THE-ART X-RAY DETECTORS FABRICATED ON PCG GROWN MERCURIC IODIDE PLATELETS.

M. R. SQUILLANTE, S. LIS, T. HAZLETT, G. ENTINE
Radiation Monitoring Devices, Inc., 44 Hunt St., Watertown, MA 02172

ABSTRACT

Recent developments on HgI_2 detectors fabricated from platelets grown by the Polymer Controller Growth (PCG) technique have resulted in a better understanding of this remarkable process and provide increased hope for the future of room temperature operable, high resolution x-ray detectors. The benefits of PCG are higher purity using reagent grade materials, better control of stoichiometry, rapid growth, and simplified fabrication. This latter benefit is even more significant when the extreme fragility of HgI_2 crystals is considered. The problems of device resolution crystal stability and uniformity, and control of platelet growth were investigated under a NASA funded program. HgI_2 devices fabricated from PCG platelets gave room temperature energy resolutions under 400 eV (FWHM) for the 5.9 eV Mn x-ray. Models for the polymer assisted transport are discussed along with an analysis of spectra obtained from PCG grown devices. Reproducibility has been demonstrated both for platelet growth and device performance.

INTRODUCTION

For many years HgI_2 has appeared to be an ideal choice for use as a room temperature X- and gamma ray detector. However, up to now the use of HgI_2 devices has been limited by lack of reproducibility of high material quality and poor device stability. Fortunately a new vapor transport technique invented by S. P. Faile[1] at Purdue University has been shown to give HgI_2 crystals of unsurpassed quality and purity. Research has been carried out to investigate and improve this promising technique.

The new technique, the Polymer Controlled Growth method (PCG), incorporates the use of an organic polymer (e.g., polyethylene) during growth which causes HgI_2 to grow as thin platelets, in a matter of a few days, rather than as tetragonal crystals. These platelets can be used directly for detector fabrication eliminating the need for crystal cleaving or shaping which can severely damage the crystal[2]. Our efforts have resulted in the reproduible growth of platelets with dimensions suitable for x-ray detector fabrication.

EXPERIMENTAL

The polymer Controlled Growth method of crystal growth is a fast and simple closed-tube vapor transport technique. However, it is fundamently very different from previous vapor transport growth techniques used with HgI_2

Mat. Res. Soc. Symp. Proc. Vol. 16 (1983) © Elsevier Science Publishing Co., Inc.

because of the addition of an organic polymeric agent to the starting materials which contributes to material purification, transport, and platelet formation.

The basic PCG process simply involves sealing reagent grade HgI_2 (99.9% pure) in a quartz ampoule with 1% by weight of a polymeric material such as polystyrene or polyethylene. Simple vapor transport can be initiated by heating one end of the ampoule to 130°C while the other end is below 130°C. Figure 1 shows the schematic of the growth furnace and temperature profile. The resultant crystals formed by tranport are primarily in the form of small platelets up to 500 microns thick with thinner platelets often reaching sizes in excess of 1 cm^2. The electrical properties of x-ray detectors made from these crystals are impressive. They have achieved the highest resolution yet observed for a room temperature x-ray detector and the values of mobility-lifetime product for both electrons and holes are among the highest reported for HgI_2, greater than 10^{-3} cm^2/V and 10^{-5} cm^2/V respectively. In addition the lowest value of the Fano Factor ever observed for HgI_2 (0.19) was reported for platelets grown by the PCG process(3). The Fano Factor is an estimate of the charge generation statistics in a material. The experimental value of the Fano Factor is strongly influenced by the material properties of the crystal; a lower value indicates higher quality material.

Devices were fabricated on platelets as removed from the growth ampoule with no additional surface etch or treatment. The fabrication technique was similar to that described by S.P. Faile, et.al[1]. Carbon contacts were painted onto the crystals ("eccocoat" or "aquadag") and thin wires embedded into the contact material as it dried. The devices were then mounted on Humiseal coated glass substrates for easy handling.

RESULTS

One of the most severe problems with any crystal growth process is final crystal purity. Impurity problems can limit performance, especially when the application places such stringent requirements on the perfection of the material as does radiation detection. HgI_2 crystals produced by previous research groups have had a history of erratic yield and lack of control over the final product. The causes of this problem may be related to the nature of HgI_2 itself. The crystal structure of HgI_2 is laminar with a wide spacing between the crystal planes allowing for ready incorporation of impurities. In addition, iodine and mercury-iodide complexes which are commonly formed with impurities can be volatile. The combination of these two factors does not allow for the efficient elimination of impurities during growth.

The principal difference between PCG and other vapor growth techniques is the addition of an organic polymeric agent. This dramatically improves the impurity content as shown in the SIMS data in Fig. 2. Impurity levels are as much as three orders of magnitude lower in the PCG platelets than in typical crystals. This is presumably due in part to gettering of the impurities during growth which eliminates the need for the 30 day multiple sublimation purification inherent in some approaches; furthermore, the organic polymer fosters the growth of thin, uniform, large area platelets. Platelet growth is not observed without the addition of these polymeric agents. The entire growth process takes less than one week and often only two days and starting materials of only moderate purity (99.9%) have been used with great success.

Table I shows the effect produced by changing the amount of polyethylene added during growth. This indicates that the polyethylene does more than just purify the starting material. The polymer or, more likely a decomposition product of the polymer, must be involved in the transport or growth of the platelets. This is supported by the fact that the rate of transport of material in the ampoule is 50 to 75% faster with polymer added than without. The type of polymer used also affects the results as shown in Table II. The behavior of various polymers has enabled us to increase our understanding of the PCG process. The difference between high-density and low-density polyethylene, for example, seems to indicate that the polymer, or a degradation product of the polymer, is acting as a transport agent and is not merely purifying the HgI_2. If the function of the polymer was merely to purify the material, the high density polyethylene, and also the other polymers, should have sufficed. In addition, if the polymer was not involved in the transport of the HgI_2, the choice of polymer would not have affected the rate of transport.

The physical properties of these platelets were examined. Phase transition temperatures were measured for various PCG crystals using a Laboratory Devices, Inc. "Mel-Temp" apparatus and an Extech digital thermometer. The temperature was raised at a rate of about one degree centigrade per minute. The transition temperature of PCG platelets was typically above 133°C. This is an indication of high purity and a very close approach to stoichiometry. Crystal defects were estimated by preferential surface etches. Using a solution of HF in HNO_3, etch pit densities of below 10^3 to above 10^6 were observed. Defect concentration appears to be related to growth temperature and rate of cooling after growth. However, no close correlation was found between etch pit density determined using the HF solution and device performance.

The electrical properties of PCG HgI_2 platelets have also been examined. HgI_2 platelets grown by PCG were used to fabricate low energy x-ray detectors by several other research teams including one at the Medical Imaging Science Group at the University of Southern California and one at the Center for Space Research at the Massachusetts Institute of Technolgy. Figure 3 shows a spectrum obtained using a device fabricated on one of our PCG platelets.

Mobility-lifetime products have also been estimated. Values greater than 10^{-3} cm^2/V for electrons and greater than 10^{-5} cm^2/V for holes have been obtained. These values were obtained using the Hecht relation for single carrier collection by measuring peak position as a function of bias voltage.

DISCUSSION

Adding polyethylene to the growth ampoule results in higher purity, better stoichiometry, platelet formation and accelerated transport. Therefore it is apparent that the polymer does enter into the transport and growth of the crystals. We hypothysize that a small organic decompositon product of the polymer (possibly even monomeric) reacts with either the solid HgI_2 or with HgI_2 vapor and thus causes faster sublimation of the source material. A possible reaction scheme is shown in equation 1.

$$HgI_2 + \,\rangle C = C \langle\, \rightleftarrows IHg - \rangle C - C \langle - I \qquad (1)$$

Since a complex is transported the stoichiometry is retained. In addition, the shape of the complex could be sterically responsible for perferential platelet growth.

The energy resolution of a HgI_2 detector can be calculated using equation 2.

$$\Delta E=[(\Delta E_p)^2+(\Delta E_i)^2+(\Delta E_{cs})^2+(\Delta E_{cf})^2+(\Delta E_{col})^2+(\Delta E_F)^2]^{1/2} \quad (2)$$

ΔE_p is the amplifier contribution to the noise, ΔE_i is the current contribution, ΔE_{cs} is due to the device capacitance, ΔE_{cf} is the 1/f noise, E_{col} is the contribution due to incomplete collection and ΔE_F is the statistical contribution. ΔE_i, $\Delta E_c S$ and $\Delta E_{1/f}$ can be obtained directly from a series of charts prepared by A. Dabrowski, et.al.[4] E_p is determined from the pulse width measured with no detector in place. ΔE_F is given by equation 3.

$$\Delta E_F = 2.355(F \epsilon E_\gamma)^{1/2} \quad (3)$$

where F is the Fano Factor, ϵ is the energy required to generate a charge pair (4.2 eV in HgI_2) and E_γ is the photon energy. The dominant terms appear to be E_i and Ecol. Figure 4 shows the change in FWHM as bias increases. Resolution improves rapidly as collection efficiency approaches 1. FWHM reaches a minimum at the point where current noise starts to dominate. At this point FWHM and pulser width both increase steadily as the leakage current increases. Present devices are limited in size because of high capacitance resulting from the thinness of the platelets. As improvements result in the growth of thicker crystals, larger area sensors will be possible. Such devices will be ideal for room temperature x-ray fluorescence and optical applications.

CONCLUSION

Polymer Controlled Growth of HgI_2 results in high purity platelets suitable for high resolution, low energy, room temperature x-ray detector fabrication using reagent grade starting materials. The crystals exhibit excellent material and electrical properties. Our investigation strongly indicates that the polymer is directly involved in both the transport and the growth of the platelets. Additional work must be done to grow larger, thicker platelets for use as gamma detectors. The Polymer Controlled Growth technique has been successfully used on HgI_2, however this research has merely scratched the surface of the capabilities of this unique process. Additional research is needed to investigate the full potential of this new growth method on a variety of important semiconductor materials. It appears likely that this process can be applied to a number of materials including other mercury compounds and semiconductors such as ZnS and ZnSe by choosing appropriate conditions and polymer. Further research will develop the PCG process into a powerful tool for the crystal grower and material scientist.

ACKNOWLEDGEMENTS

We wish to thank Gerald Huth, Andrzej Dabrowski, Jan Iwanczyk, Jan Checkinsky and Jeff Barton at USC and George Ricker, Allan Warren, and John Vallerga at MIT for their helpful discussions and measurements on HgI_2 platelets and detectors. We are especially grateful to Sam Faile for his many helpful discussions and insights concerning the PCG process. This work was supported by NASA under contract No. NASW-3437.

REFERENCES

1. S.P. Faile, A.J. Dabrowski, G.C. Huth, J. S. Iwanczck J. Cryst. Growth. 50, 50, 752 (1980).

2. W.B. Yelon, R.W. Alkire, M.M. Schieber, L. van den Berg, S.E. Rasmussen, H. Christensen, J.R. Schneider, J. Appl. Phys. 52, 4604 (1981).

3. G.R. Ricker, J.V. Vallerga, A.J. Dabrowski, J.S. Iwanczyk, G. Entine, Rev. Sci. Inst. 53, 700 (1982).

4. A.J. Dabrowski, G.C. Huth, IEEE Trans. Nuc. Sci. NS-23, 102 (1976).

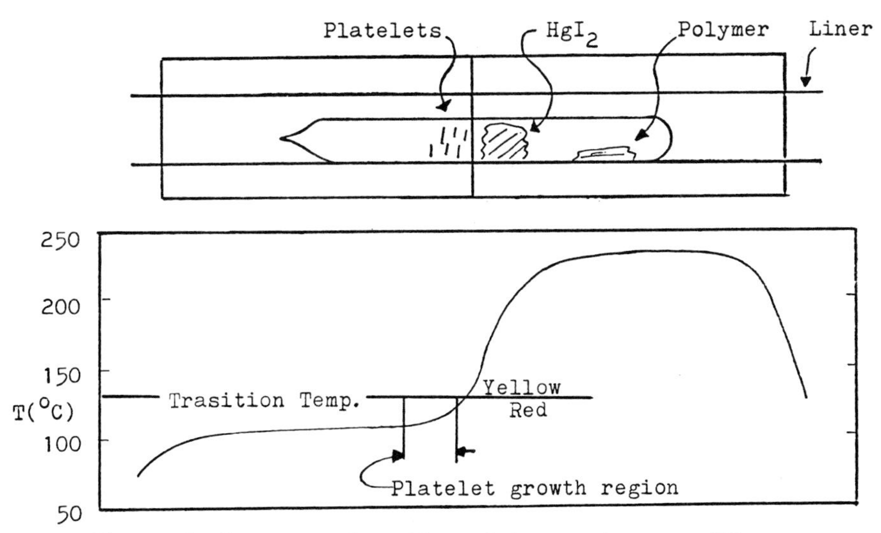

Figure 1. Furnace schematic and temperature profile

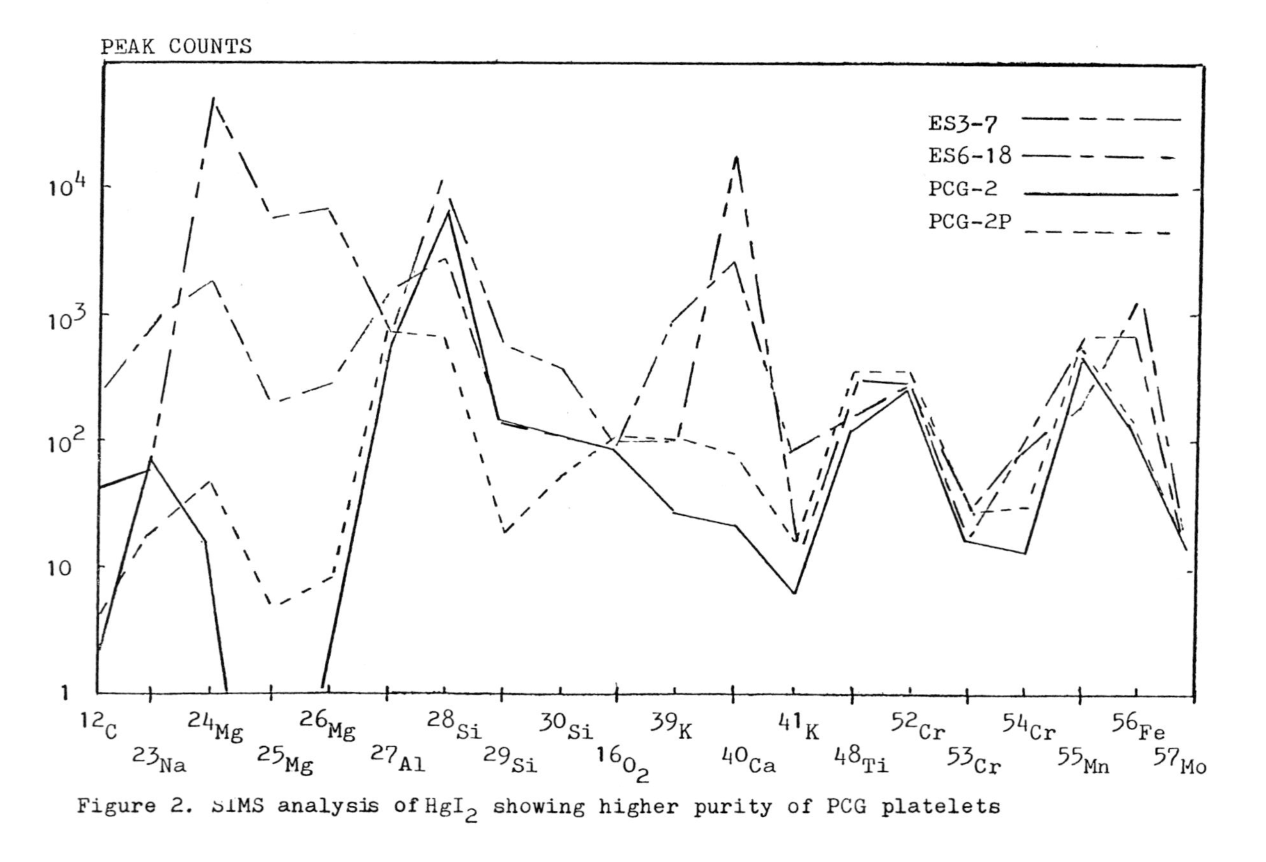

Figure 2. SIMS analysis of HgI_2 showing higher purity of PCG platelets

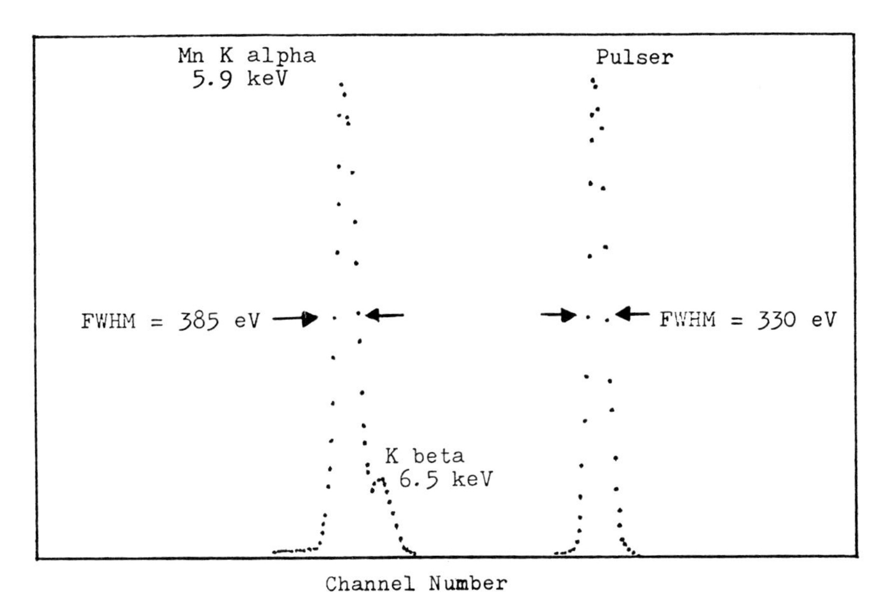

Figure 3. Spectrum from Fe-55 source using PCG platelet

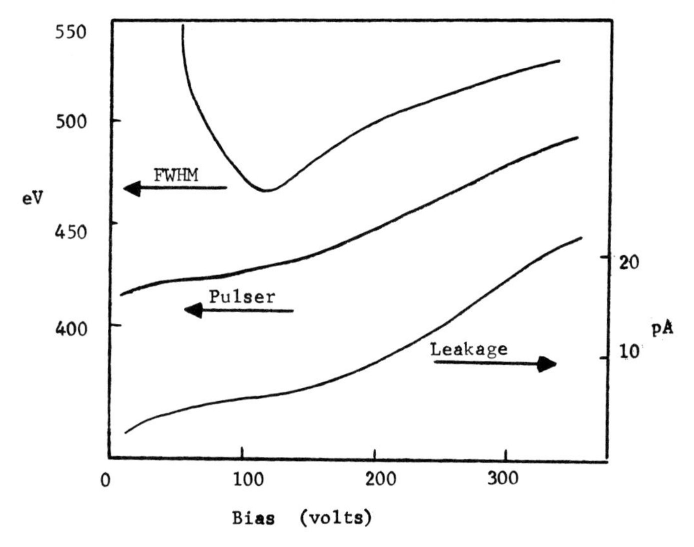

Figure 4. Effect of increasing bias voltage

Table I
Effect of Varying the Amount of Polyethylene Transport Agent on Platelet Growth

% Polymer (by weight)	Comments
0%	No platelets, tetragonal crystals
0.1%	Few small platelets, tetragonal crystals
1.0%	Well shaped platelets and tetragonal crystals
3.0%	Large thin platelets, small tetragonal crystals.

Table II
Effect of Polymer Type on PCG Platelet Growth

Polymer	Devices	Comments
Polyethylene		
Low density	Good	Standard Procedure
High density	No devices	Slow transport, tiny plates
Polystyrene	Good	Similar to polyethylene
Styrene	Good	Similar to polyethylene
Polyvinyl chloride	No devices	Slow transport, poor plates
Polymethyl-Methacrylate	No devices	V slow transport, no plates

THE FABRICATION OF MERCURIC IODIDE DETECTORS FOR USE IN WAVELENGTH DISPERSIVE X-RAY ANALYSERS AND BACKSCATTER PHOTON MEASUREMENTS

John H. Howes and John Watling
AERE HARWELL, Didcot, Oxon. OX11 ORA, UK.

ABSTRACT

This paper describes the fabrication of mercuric iodide nuclear radiation detectors suitable for X and gamma ray spectrometry at room temperature. The active area of the detectors studied are between 0.2 and 1.5cm sq and they are up to 0.5mm thick. The method of producing a stable electrical contact to the crystal using sputtered germanium has been studied. The X-ray resolution of a 1.5cm sq. area detector at 32 keV is 2.3 keV FWHM when operated at room temperature in conjunction with a time variant filter amplifier. A factor which is important in the fabrication of the detector is the surface passivation necessary to achieve a useful detector life.

This type of detector has been used on a wavelength dispersive X-ray spectrometer for energy measurements between 10 and 100 keV. The advantages over the scintillation counter, more commonly used, is the improved resolution of the HgI_2 detector and its smaller size. The analyser is primarily used for the detection of low levels of heavy metals on particulate filters. The detectors have also been used on an experimental basis for gamma ray backscatter measurements in the medical field.

INTRODUCTION

Interest in mercuric iodide (HgI_2) for radiation detection started at the turn of the century when its photosensitivity was observed, a particular interest was shown in using small crystallites in photographic emulsions. The application related to photographic emulsions did not develop very far on account of the instability of HgI_2 with time. In the 1950's a renewed interest in single crystal HgI_2 was shown and several researchers investigated the optical and photoelectrical properties of the material.

At room temperature HgI_2 crystals are of the red tetragonal form, above 127°C a phase transformation occurs to a yellow orthorhombic habit. This transformation is a shift of the <100> plane in the tetragonal form to the <110> plane in the orthorhombic form, the <001> plane of the crystal is preserved in the phase transformation. The phase-transformation kinetics was described by Newkirk in 1956(1). Associated with the change in crystal habit at 127°C is a change in photosensitivity, the orthorhombic form being about 0.1% that of the tetragonal form(2). The principal interest in electrical and photoelectrical properties of HgI_2 has been in the tetragonal crystal form. Bube(3) studied several properties of the material, notably location and temperature dependence of the absorption edge by measurement of optical transmission, reflectivity and photoconductivity spectra. He also investigated the temperature dependence of dark current for various samples of tetra-HgI_2 produced by solution regrowth, vapour phase growth, and melted and recrystallised layers.

The interest in HgI_2 as a nuclear radiation detector started in 1971 with Willig and Roth(4) showing that the properties of red HgI_2 had potential as a detector of X and gamma radiation. With a band-gap of 2.1eV and low hole-electron pair creation energy of 4.2eV at 300^oK, its use as a room temperature X and gamma ray detector has many operational advantages over other available types of detector. A number of groups (5, 6, 7 and 8) showed that HgI_2 with its high atomic number (80-53) was particularly suitable for X and gamma ray spectrometry at room temperature. Since those early developments up to 1975, there has been a steady advance in the understanding of the physical and electrical properties of the material, in particular the various methods of crystal growth. Single crystals of HgI_2 can be grown from vapour phase or by solution regrowth, both methods can result in HgI_2 crystals suitable for successful detector fabrication. The solution regrowth method has been used to grow single crystals of up to 100g. Single crystals are grown between 25 and 40^oC through the temperature-driven shift of the reversible chemical equilibrium of complexes of HgI_2 with dimethylsulfoxide. This method developed by Nicolau and Joly(9) allows corrections in stoichiometric deviation during growth. The start material requires purification by vacuum sublimation prior to crystal growth. Crystal growth by vapour phase falls into three different methods. The first, is the static temperature profile method,(5). Secondly, the temperature oscillation method which was initially developed by Scholz(10) which involves a periodic reversal of the temperature gradient between source material and the crystal, resulting in alternating crystal growth and re-evaporation. This method has been further refined by Van der Berg(11) by the periodic oscillation of the source material temperature and the periodic oscillation of the crystal temperature. The third method is described by Faile(12); platelets of HgI_2 are grown in a quartz tube involving the use of an organic polymeric material such as polyethylene or styrene. With careful control of the temperature gradient down the quartz tube it is possible to produce small crystals of a few mm^2 area and up to 0.4mm thick. The resulting crystals or platelets have been shown(13) to produce small X-ray detectors capable of achieving a resolution of better than 300eV FWHM on MnKα (5.9keV), which is comparable to HgI_2 grown by other vapour transport methods.

DETECTOR FABRICATION

Various techniques have been used for cutting and shaping HgI_2 into thin slices suitable for making detectors. One method is to mechanically cleave the crystal along the <001> crystal plane. This method can result in high dislocation damage along the cleaved faces which can influence the detector performance in terms of leakage current, requiring the etching of up to 300µm from each <001> face of the crystal. There is evidence(6,14) that mechanical cleaving creates trapping levels at 0.2 and 0.25 eV as determined by thermally stimulated conduction measurements. Another method of cutting uses a reciprocating filament of wire or cotton with an aqueous solution of 20% KI. This method depends on chemical mechanical cutting and it is necessary to protect the surfaces from the KI solution which tends to dissolve all exposed areas of the crystal. After cutting, a methanol etch will reveal any dislocations. Polishing and etching with a KI solution on

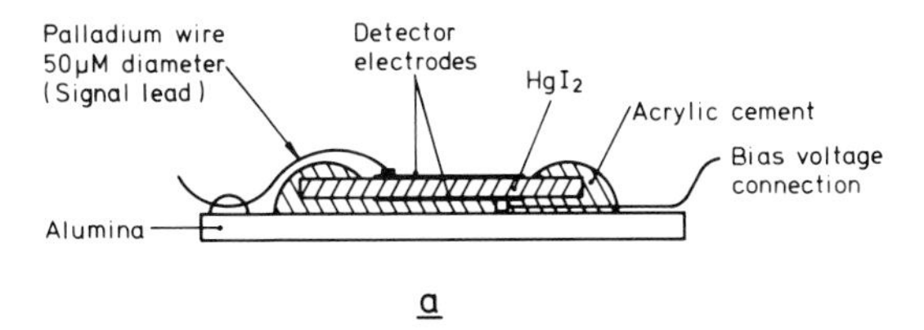

a

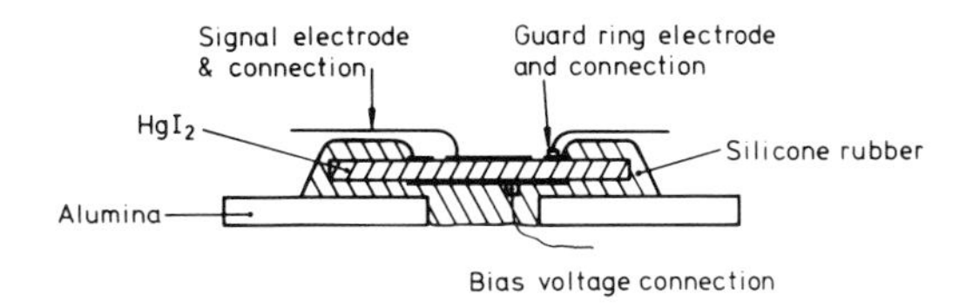

b

Fig. 1. Methods of mounting HgI_2 crystals:-
(a) Using acrylic cement and
(b) Detector guard-ring using silicone rubber.

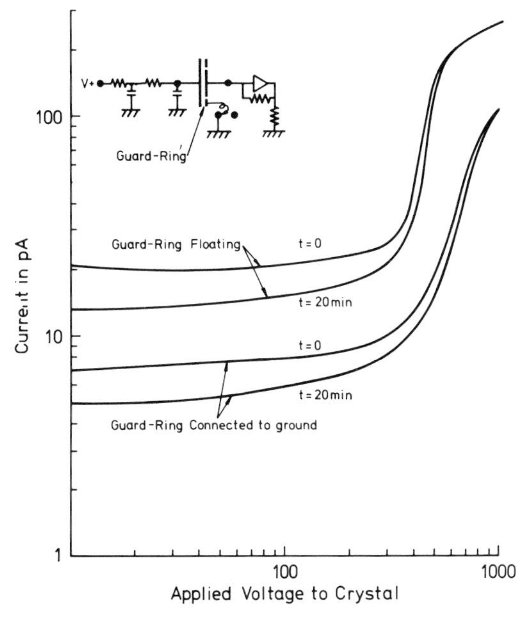

Fig. 2. Current-voltage relationship of HgI_2 crystal with guard-ring floating and grounded, measurement of leakage current at times t = 0 and t = 20 mins. Active area of detector $30mm^2$, $350\mu m$ thick.

a fine felt pad results in a suitable surface to deposit electrodes.

Methods of deposition of conducting electrodes onto HgI_2 are limited on account of chemical reactions with the crystal; the high vapour pressure of HgI_2 causes the preferential loss of iodine at the crystal surface. Contacts(15) using carbon, palladium and germanium have been used by many researchers. Carbon is deposited from an aqueous or alcohol solution using a fine air brush. Palladium and germanium contacts can be successfully deposited by using thermal heater evaporation in a system where the exposure to low pressure is limited to a few minutes. A thermal shutter is used to mask the crystal from the heater prior to the evaporation cycle. The thickness of the Ge and Pd contacts are between 150 and 300 Angstroms. Gold evaporated films are not satisfactory because the gold reacts with the mercury forming an amalgam.

Encapsulation of the HgI_2 is essential if the detector is to have a reasonable operational life and this is achieved by using either acrylic cement Fig.1a or silicone rubber as shown in Fig.1b. A further method of overcoming degradation in detector peformance is to use a sealed enclosure filled with dry nitrogen. In our experience detectors with good charge collection encapsulated in acrylic cement and stored in a dry atmosphere have shown stable operation over two years. Degradation of leakage current has been noticed on crystals which have not been encapsulated. On the edges of the <001> plane, a change of colouration occurs from a clear red to dull red. This maybe due to interaction with water vapour in the atmosphere.

DETECTOR CHARACTERISTICS

Short term changes in detector leakage current at a given voltage are indication of the performance of the detector as a stable X or gamma-ray spectrometer. Using these changes in leakage current together with collimated X-ray measurements it is possible to obtain an indication of the type of detector defect, e.g. crystal morphology, mechanically induced damage or electrode contact defects. The change in detector leakage current versus the applied voltage for the case of guard ring connected and guard-ring floating is shown in Fig. 2. This change in leakage current 20 mins after the application of the polarising voltage is typical of detectors of good spectrometry grade. In the case of poor detector performance the leakage current change with time is very much greater, also the current is often erratic indicating a dielectric breakdown phenomena.

The X-ray resolution of the detector shown in Fig. 2. is 2.5keV FWHM (full width half maximum) at 59.6 keV with the guard ring grounded, with a floating guard-ring the resolution degrades to 3.9keV FWHM. This degradation is not accounted for in terms of increased leakage current alone but is more the effect of incomplete charge collection at the periphery of the signal electrode, due to a region of weak electric field. The Am^{241} spectra shown in Fig. 3. shows the difference in charge collection with and without the guard ring connected. When using a floating guard ring at higher counting rates the detector performance is more rapidly degraded. The difference in spectral distribution between Fig.3a and b is accounted for in the case of the floating guard-ring by incomplete charge collection at the periphery of the collector electrode. This poor charge collection is due to the region of weak electric field where the carriers generated will have low velocity. The continuum background is smaller when the guard-ring is connected to the same

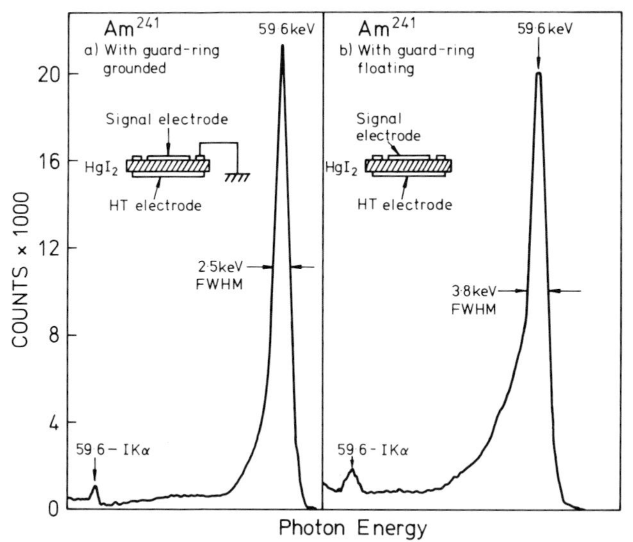

Fig. 3. Spectra obtained with HgI_2 detector using Am^{241}
(a) With guard-ring ground and
(b) Guard-ring floating.

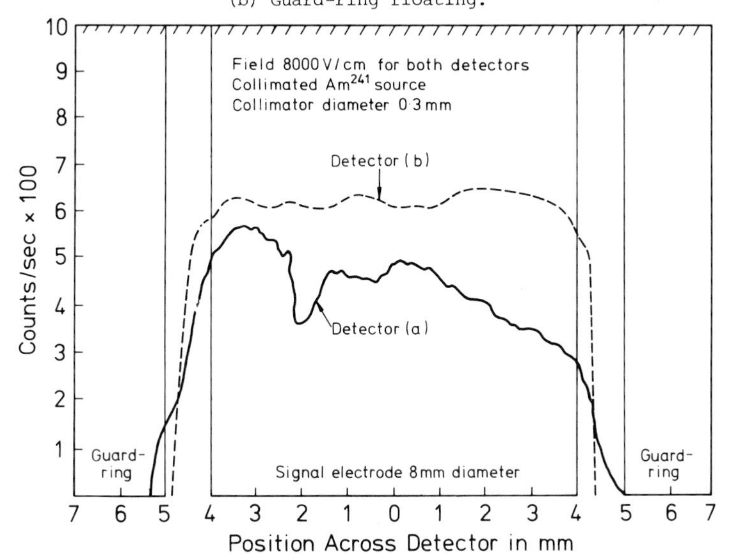

Fig. 4. Sensitivity scan across 8mm diameter HgI_2 detectors with 2mm guard-rings, showing the response to Am^{241} (59.6keV) through a 0.3mm dia. collimator.

potential as the signal electrode. Charge collection deficit arising from electric field distortion is reduced by using a grounded guard-ring geometry. Similar effects arising in germanium detectors with weak electric fields or slow charge collection have been partially eliminated electronically by using a series switch technique with a time variant filter(16) resulting in improved energy resolution. A further cause of poor energy resolution is the variation in charge collection(17) in the crystal on a micro scale. Variation in crystal homogeneity results in changes in the electric field E in the dielectric, trapping time τ and drift mobility μ. Since these parameters are related to the drift length λ through $\lambda = \mu\tau E$, variations in drift length are correlated with material inhomogeneities. An anomalous feature of HgI_2 is related to the average energy required to create an electron hole pair, ε , which generally increases with increasing energy band-gap of the crystal. Values of ε versus energy band-gap Eg for various crystals have been calculated semi-empirically by Klein(18) and fit the relationship $\varepsilon = 2.67\ Eg + 0.87eV$. The value measured for HgI_2 obtained experimentally is 4.2eV(6) which does not fit the above relationship.

The gamma-ray response of two crystals of similar geometry is shown in Fig. 4. The detectors are of circular geometry with a 2mm wide guard ring. The data is obtained by scanning the detector with a collimated Am^{241} source, using a 15keV window on the 59.6keV gamma ray line. Detector(a) exhibits poor energy resolution, 9.8keV FWHM at 59.6 keV, when the crystal is totally illuminated. If the collimated source illuminates a region away from the area of reduced sensitivity shown in curve(a), the resolution is 5.0keV FWHM. The reduction in count rate corresponds to dislocation lines in that region of the crystal. Curve b Fig. 4. is an example of a scan on a good spectroscopy grade detector with a resolution of 2.4keV FWHM for total illumination of the crystal with Am^{241}. A method of evaluating crystals reported by Schieber et al(19) is to measure the dc current change with time at X-ray dose-rates up to 400R/min. Then observe the time decay of the dc current at a constant dose-rate over several hours. This method shows changes in current of up to 30% in 4 hours in poor detectors, however; with good spectroscopy grade crystals the dc current decay is negligible at high gamma ray dose-rates.

APPLICATION OF MERCURIC IODIDE AS A DETECTOR IN WAVELENGTH DISPERSIVE X-RAY SPECTROMETER

For the past twenty years or so, scintillation detectors have been fitted to wavelength dispersive X-ray spectrometers, for the detection of energies greater than 10keV. A scintillation detector has a high quantum counting efficiency and short dead time, however its resolution is poor (4keV at 10keV). The HgI_2 detector has been shown to improve the overall resolving power of the crystal spectrometer at low angle, high energy X-ray wavelengths, thus giving greater spectral and elemental coverage. Its resolving power overcomes high order interferences making line overlap corrections unnecessary, and enables analysis to be carried out more rapidly.

Crystal spectrometers, which depend upon the Bragg equation ($n\lambda = 2d \sin\theta$) produce higher order lines ($n = 2, 3$ ----) which in certain circumstances can interfere with the lines of other elements of lower energy. For example, in the case of the determination of lead (Fig.5) using its $L\beta$ line (12.61keV) serious overlap by the second order (25.3keV) from tin $K\alpha$ occurs. The scintillation counter is unable to resolve the

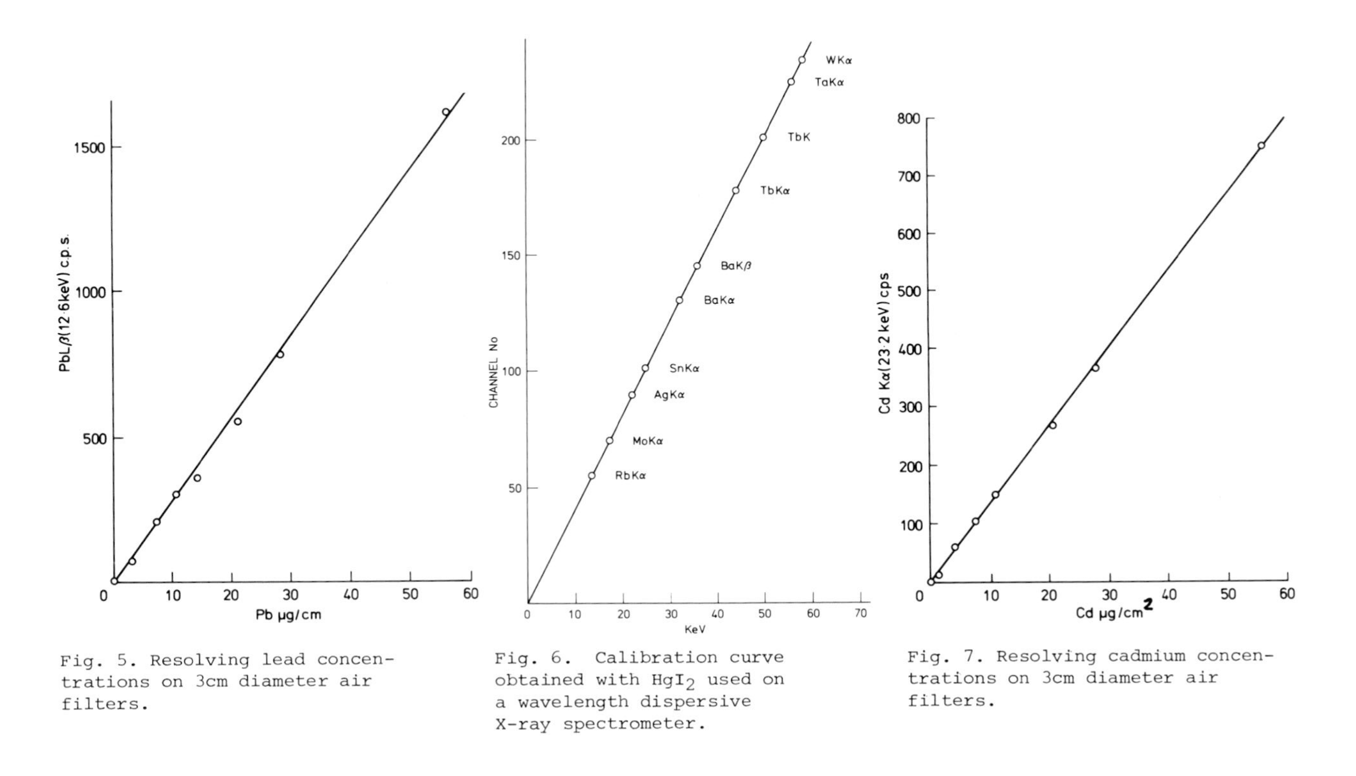

Fig. 5. Resolving lead concentrations on 3cm diameter air filters.

Fig. 6. Calibration curve obtained with HgI_2 used on a wavelength dispersive X-ray spectrometer.

Fig. 7. Resolving cadmium concentrations on 3cm diameter air filters.

peaks satisfactorily due to inadequate resolution.

On the other hand HgI_2 with its excellent resolution (1 keV at 10keV), and efficient stopping power for the high energy X-rays, provides the means to overcome this difficulty. Lead has been determined successfully in the presence of tin in air particulates trapped on air filters. Although due to geometric factors lower counting rates were observed with the HgI_2 detector, the signal to noise ratio has been found not to differ significantly from the scintillation counter, so that similar detection limits are observed for the same counting times. The calibration graph obtained for the determintion of lead in the range 0 to 60μgm/cm^2 (using 3cm diameter filters) in the presence of tin is shown in Fig. 5.

As a wavelength dispersive X-ray spectrometer scans toward a low angle; i.e. the high energy portion of the X-ray spectrum the overall resolution of the instrument decreases and the X-ray lines (energies) tend to crowd together. This can be partially overcome by using crystals of smaller d spacing, but this causes loss of sensitivity due to the poorer reflectivity of short d spacing crystals. The HgI_2 detector provides a means of improving the overall resolution of the crystal spectrometer thus increasing its applicability to the K lines of heavier elements. Fig. 6 is a calibration plot of the detector used with the wavelength dispersive X-ray spectrometer using a pulse height analyser.

Tantalum has been measured at low levels in low alloy steels using the Ta K_α (57.5keV) line and a tungsten target (59.3keV) X-ray tube. No interference from tungsten was recorded at the Ta K_α position, even when a specimen of pure tungsten was examined.

Tin (25.3 keV) and antimony (26.4 keV) have been successfully determined in low alloy steel, using the K lines rather than the L lines of lower energy. This permits a greater sample depth to be examined, which reduces problems due to possible sample heterogeneity and also reduces the need for a much finer specimen surface finish, a prerequisite when using the less penetrating X-rays.

Similarly, particulate air filters can be analysed simply with K X-rays using the so called 'thin film' technique for sample calibration. This method requires that the sample is thinner than the critical thickness (which relates to the energy of the analyte line employed) otherwise a multiplicity of standards would be needed in order to carry out matrix correction procedures, necessary when using the less penetrating L X-rays.

As an example of this analysis using HgI_2, cadmium(Fig.7) has been determined down to a lower limit of detection of 0.2μg/cm^2 in a counting time of 300 secs.

BACKSCATTER PHOTON MEASUREMENTS

In the medical physiology field there is an interest in measuring pulmonary edema(20). By using a HgI_2 detector in a conical collimator arrangement shown in Fig. 8, it is possible to measure the photon backscattered peak resulting from illuminating the subject with 150keV radiation from Tc^{99} or Co^{57}. The purpose of the measurement is to follow the time course of lung water changes in acute or chronic pulmonary edema induced by either increased microvascular permeability or increased microvascular hydrostatic pressure. The method adopted for this measurement

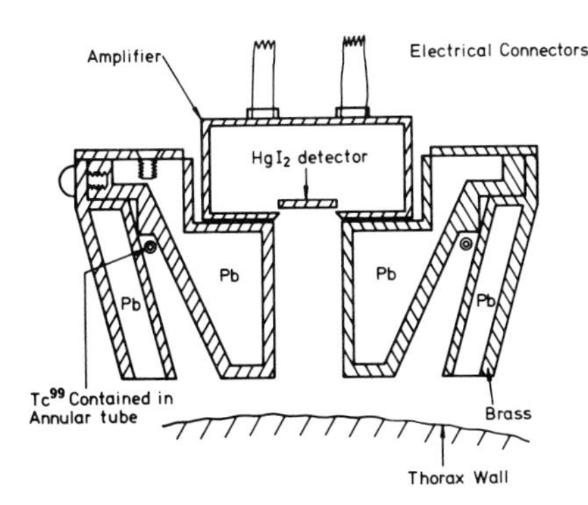

Fig. 8. Collimator arrangement for photon backscatter system using HgI_2 detector for measurement of pulmonary edema.

is shown in Fig.8. The conical collimator illuminates a region 3 to 5cms into the thoracic wall and the backscattered photons are detected by a HgI_2 crystal in the centre of the collimator. The aperture of the detector collimator can be adjusted. The advantage of the HgI_2 detector is its compactness in the collimator and its resolution in terms of the backscattered photons at the energies used. The method appears as a useful clinical technique(21) which is relatively inexpensive.

CONCLUSIONS

The advances in crystal growth of HgI_2 both by the vapour transport and solution regrowth methods have reached the stage of showing great promise for the material as a detector in X-ray fluorescence spectroscopy at room temperature. The fabrication techniques applied to the grown crystals have a considerable influence in manufacturing spectroscopy grade detectors. One of the problems to be solved is the method of encapsulation to obtain a reasonable operational life. Controlling the environment of detectors by operating in dry nitrogen show promise and is simple to implement . For photon energies above 150keV, performance is limited by hole trapping and polarising effects. Although more recently with improved crystal growth techniques these polarising effects are not so serious. A better understanding of the physics of electrical contacts to HgI_2 is required, which may then influence the higher energy gamma-ray performance of the detectors in terms of charge collection.

There are many potential X-ray spectroscopy applications which can be successfully applied with the economic benefit of room temperature operation. There are indications that in cooling HgI_2 detectors down to $250^{\circ}K$ the hole mobility improves slightly. Use of a small thermoelectric cooler (3 watts) would have the added advantage of cooling the input n-channel field effect transistor with the detector. This would result in a further improvement in low energy X-ray energy resolution at a modest cost.

ACKNOWLEDGMENTS

The authors are indebted to a number of their colleagues for their close collaboration during studies in the areas described above. Particular thanks to J. C. Woodmore, F. L. Allsworth and T. G. Jones. We would also like to acknowledge the co-operation and material help given by R. L. Lynn, W. F. Schnepple and L. Van der Berg of E. G. & G. Inc., Advanced Measurements Group, Santa Barbara.

REFERENCES

1. J. B. Newkirk, Acta Metallurgica 4, 316 (1956)

2. E. F. Gross, Zhur Tekh Fiz 25, 1661 (1955)

3. R. H. Bube, Phys. Rev. 106, No. 4, 703 (1957)

4. W. R. Willig and S. Roth. Proc. of International Sym. on CdTe, authors A. Cornel and P.Siffert, Strasbourg (1971).

5. H. L. Malm. IEEE Trans Nuc Sci NS 19 No.3 (1972) p 263.

6. J.P.Ponpon et al. IEEE Trans Nuc Sci NS22 (1975) p 182.

7. A.P.Swierkowski et al. App. Phys. Lett.23(1973) 281.

8. M. Schieber et al. IEEE Trans Nuc Sci NS21, No.6(1974) p.305.

9. I. F. Nicolau, J.P.Joly, J of Crystal Growth 48 (1980) 61.

10. M. Scholz. Acta Elektron 17 (1974) 69.

11. M. Schieber, W.F.Schnepple & L. Van der Berg. J of Crystal Growth 33(1976) 125.

12. S. P. Faile et al. J of Crystal Growth 50 (1980) 752.

13. J.P.Barton et al. Advances in X-ray Analysis, Vol 25 (1982) 30th Denver Conference.

14. Proc. of Workshop on HgI_2 Strasbourg 1975 (Ed. P. Siffert).

15. Z.H.Cho et al. IEEE Trans Nucl Sci NS22 (1975) 1229.

16. F.L.Allsworth et al. Nuc. Inst & Meths. 193 (1982) 57.

17. A. J. Dabrowski. Advances in X-ray Analysis Vol 25 (1982) 1.

18. C. A. Klien, J. App Phys. Vol 39 (1968) 2029.

19. M. Schieber et al. IEEE Trans Nucl Sci. NS24 (1977) 148.

20. D. S. Simon et al. Journal of Applied Physiology Dec 1979. 1228.

-SEARCH FOR A CORRELATION BETWEEN STOICHIOMETRY, TSC SPECTRA AND -NUCLEAR SPECTROSCOPIC CAPABILITIES OF HGI_2.

A.TADJINE, D.GOSSELIN, J.M.KOEBEL and P.SIFFERT
CENTRE DE RECHERCHES NUCLEAIRES, Laboratoire PHASE
67037 STRASBOURG CEDEX France

ABSTRACT

Mercuric iodide is, potentially, an interesting material for room temperature high efficiency gamma or X-ray spectroscopy. However, still several parameters limit the performance of these devices and even the crystal growth is not fully under control today, poor reproductibility being reached. An optimization of growth conditions seems difficult since no parameter of the material could be correlated with the spectrometer's performance. Therefore, the effective progress in recent years was rather limited.

The goal of this paper is to investigate systematically if a correlation can be found between bulk stoichiometry for crystal grown under certain conditions, thermally stimulated current (TSC) measurements and nuclear performance. In this study, two families of crystals could be identified. Best results are obtained for nearly stoichiometric materials, exhibiting a specific TSC spectrum. In particular, the peak appearing at 173 K constitutes a good indication of the detection capabilities of the material, both for spectroscopic resolution and polarization.

The growth conditions have been adjusted to produce in a fully reproducible way high performance devices.

INTRODUCTION

Due to its high atomic numbers (80-53), to its large bandgap (2.15 eV) and to the low energy needed to generate one electron-hole pair, mercuric iodide is potentially an interesting material for the detection of γ and X -rays at room temperature. However, the performance of these devices are non-reproductible and limited by the presence of trapping centers, which reduce both lifetime and transport properties. To explain the presence of these defects, several explanations can be considered : presence of chemical impurities or physical defects, stoichiometric deviations.

Recently, it has been shown in the literature, by cristallographic studies (1)chemical analysis (2) and Rutherford backscattering (3), that strong stoichiometry departures may exist in the bulk as well as in closer surface vicinity.

Mat. Res. Soc. Symp. Proc. Vol. 16 (1983) © Elsevier Science Publishing Co., Inc.

The goal of this paper is to correlate the presence of trap levels with stoichiometric deviations in crystals prepared under various conditions.

The trap levels were measured by the thermally stimulated current method (TSC), whereas the volumic stoichiometry was studied by chemical complexometry analysis. Finally, the results were compared to detection properties.

I. CRYSTAL GROWTH AND SAMPLE PREPARATION

The crystals used in this study were grown by the temperature oscillating method (TOM), with a vertical furnace, as developed by SCHOLZ (4). In a first step of the growth process, the crystal temperature oscillates around 120°C, this value is the upper limit before the occurence of a phase transition. Once the crystal has reached a volume of about 0.5cm^3 and once the risk of spontaneous nucleation is low enough, the growth temperature is kept constant. The final crystal weightes between 50 and 60 g.

The volumetric stoichiometry measurements are performed on cleaved samples of the as-grown crystal. For TSC measurements and nuclear spectrometry experiments, 1 to 1.5 mm thick samples are cut with a wire saw by using KI solution along (001)crystal planes. Finally, the samples are polished on wetted felts with a 10% KI solution. It should be noticed that the concentration of KI, as well as the sawing speed are important parameters with respect to the smooth surface aspect. After polishing, the samples are etched in methanol or acetone and their final thickness is reduced to about 700μm.

It is interesting to notice that this procedure allows low leakage current structures manufacturing, but does not improve the detector's performance. In our opinion, the surface preparation step is not the most critical, since good spectrometers can also bee realized by crude cleaving of the crystals (5). Finally, the preparation of the detector is achieved by painting contacts 3-4 mm in diameter on both sides of the samples with eccocoat 257, the device is embedded by a spray of KF electronic isolant.

II. TSC MEASUREMENTS

The principles of the TSC measurements are well known (6) and are no longer described here. This method is well suited for the detection of deep traps in high resistivity materials (7) and has already been used by several authors (6, 8-13) for investigating HgI_2.

The samples are mounted in a special cryostat allowing a linear increase of the temperature between 80 and 300 K at a rate of 0.4 K/s. This temperature is recorded by means of a platinum resistor embedded in the holder, very close to the device under investigation. The latter is stuck to the holder by eccocoat, to ensure a good electrical and thermal contact. Bias voltages of 20 or 40 V are applied during both heating and cooling cycles.

After cooling to 77 K, the traps are filled with interband light (0.54 μm) during about 5 min. Care must be given to avoid any change of the material when kept for longer periods of time under vacuum with illumination and fast thermal cycling. Here, we used slow cooling cycles of about

TABLE I

Maxima in the TSC curves and corresponding energy

Maxima in TSC spectra (° K)	100	110	125	155	173	205	225
Corresponding energy level (eV)	0.16	0.28	0.32	0.40	0.45	0.50	0.59

TABLE II

Relative intensities of the TSC peaks

Crystal / T_{max} (°k)	100	110	125 to 132	155 to 160	173 to 180	205	225
SO1			475	800	1925		
SO2		175	200	600	1000	405	175
SO3		400		1475	5930	1075	
S11	25		50	109	150	119	
S12	not well def.		90	160	230	60	50
S14	87		162	612	4500		
S15		100	90	85	140		60
S16		30	55		140		5
S17		110	100		210		20
S18			130	170	200		10

TABLE III

Stoichiometry of various crystals

Material	Element	Nr of moles x10E-3	Stoichiometry	error °/°°
Starting material Backer	Iodine	1.7585	2.006	2.4
	Mercury	0.8767		
SO3 crystal	Iodine	1.7808	2.048	0.9
	Mercury	0.8694		
S17 crystal	Iodine	1.7628	2.002	0.7
	Mercury	0.8802		

3h. Some samples have been stabilized by room temperature illumination for long periods of time (14).

In general, the same peaks are observed in all TSC spectra but their intensities (relative and absolute) are quite changing from sample to sample. The temperature of the maxima and the corresponding energies calculated with a deconvolution program, based on the method of GARLICK and GIBSON (15)are reported on Table I. (The energy corresponding to the last peak at 225 K has been evaluated by STUCK et al (16) by various methods). The intensities of the lines measured on various crystals of nearly same active volume are reported on Table II.

Two groups of crystals are found with respect to the level at 173 K:
- in the first one (crystals SO1, SO2, SO3, S14) the intensity of the level at 0.45 eV lies between 1000 and 6000 pA, whereas in the second group (S16, S17, S18, S11, S12, S15) the intensity is about 100-200 pA. These results will be discussed latter, in correlation with the stoichiometry analysis and detector performance.

III. STOICHIOMETRY ANALYSIS

The chemical analysis of HgI_2 has been performed by complexometry of HgI_2 and determination of Hg and I concentration.

A. For iodine, the concentration was determined by oxidation with potassium iodate, by a method which has been developed as soon as 1903 (17) and which is described extensively in "Reagent chemicals and Standards" (18-20). DISHON et al (2) used already this approach for the same purpose.

In presence of chlorhydric acid at sufficient concentration ($\geq 4N$) the oxidation reaction can be written as :

$$2\ KI + KIO_3 + 6\ HCl \rightleftarrows 3\ KCl + 3\ ICl + 3H_2O.$$

To detect iodine, we use chloroform i.e. the change of color from violet to colorless at the equivalent point of the titration.

B. The concentration of mercury was done through complexometric titration (2, 21, 22) with ethylenediaminetetra-acetic acid (EDTA). The end of the titration is determined with an indicator which is sensitive to the presence of metallic cations (black erichrome T). It should be noticed that problems can arise in this titration, due to the existance of I^- anions which act as masking agents of Hg(II)(23). Therefore, one has to oxide HgI_2 in order to eliminate iodine and iodide. The procedure we used is slightly different from that reported in ref. 2, since it was possible to titrate in a reproducible way a greater amount of mercury by an addition of NaOH, instead of NH_3 solution. In the NH_3 case, reproductible measurements can only be achieved with the same volumes of acid and NH_3, due to the diminution of the stability constants for the metal Erio T and for the metal EDTA complexes, with increasing values of NH_3 and NH_4^+ as shown in Fig. 1 (24).

On Table III we have reported the results we obtained first on the Backer starting material, then on two crystals typical, from the TSC results point of view.

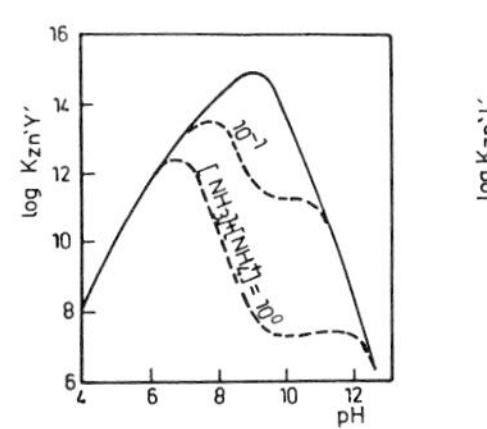

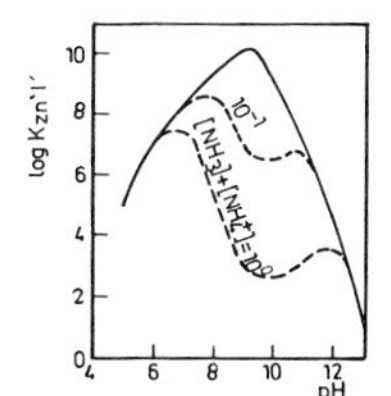

Figs. 1 a and b :conditional stability constant K_{Zn} "Y" of the Zn EDTA complex as a function of pH for various values of $[HNO_3] + [NH_4^+]$ idem for Zn Erio T complex

It appears that the values of the starting material are close to that of DISHON et al (2). On crystal S17 the stoichiometry is nearly ideal, whereas large deviations are seen in sample SO3.

IV. NUCLEAR DETECTION RESULTS

The spectroscopic properties of the detectors were tested with ^{55}Fe γ-rays (5.9 keV), by irradiation through the negative electrode, to enhance the electron charge collection. The best results are obtained for the crystal S17 as shown on Fig. 2. For this category of crystals no polarization was observed. However the resolution of the detectors is better for low energies (^{55}Fe) and degradates for the 60 keV ^{241}Am line.

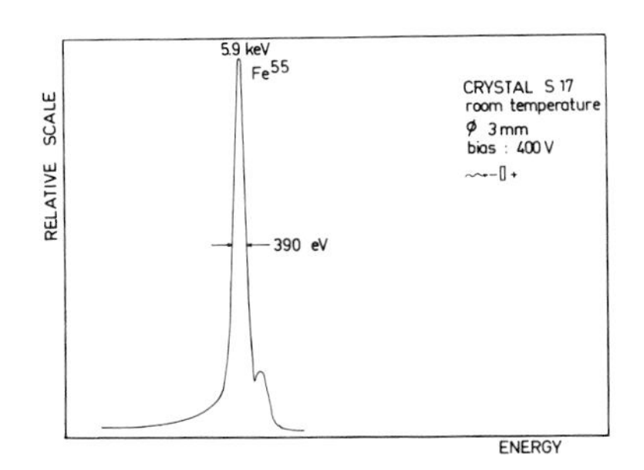

Fig. 2: the best resolution for the 5.9 keV γ - line of ^{55}Fe obtained with a HgI_2 detector 3 mm in diameter operating at room temperature at a bias voltage of 400 Volts.

The poorest performance were observed on SO3, which exhibited fast polarization, leading to the spectrum vanishing after 5 min (Fig. 3). On Table IV we summarized the behaviour of all crystals with respect to polarization.

V. DISCUSSION AND CONCLUSION

-If one compares the TSC spectra to the polarization behaviour, it appears that all crystals of the first family (SO1, SO2, SO3, S14), which show intense TSC peak at 173 K, exhibit fast polarization. It is, therefore, highly probable that the level at 0.45 eV relates directly or indirectly to polarization. The same category of crystals shows as strong stoichiometric deviation. (Fig. 4).

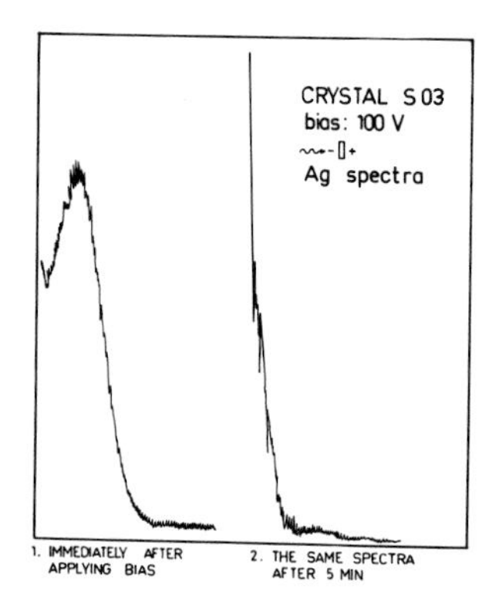

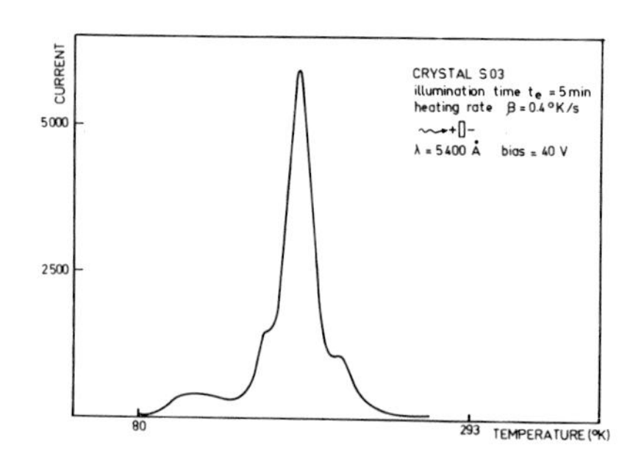

Fig. 4 : typical TSC spectrum of the first crystal family

Fig. 3: typical polarization effect on SO3 crystal under Ag X-ray excitation, immediately after bias switched on and after 5 min.

-All crystals of the second family (S16, S17, S18), which have a low intensity peak at 173 K can be employed to prepare good detectors. These crystals are nearly stoichiometric. It appears also that S16, S17 and S18 show low intensity peaks at 205 and 225 K when compared to the 173 K peak, whereas S11, S12 and S15 have peaks at 155, 205 and 225 K more pronounced. However, this later can also been used for spectrometer realization with only slight polarization. (Fig. 5 to 7)

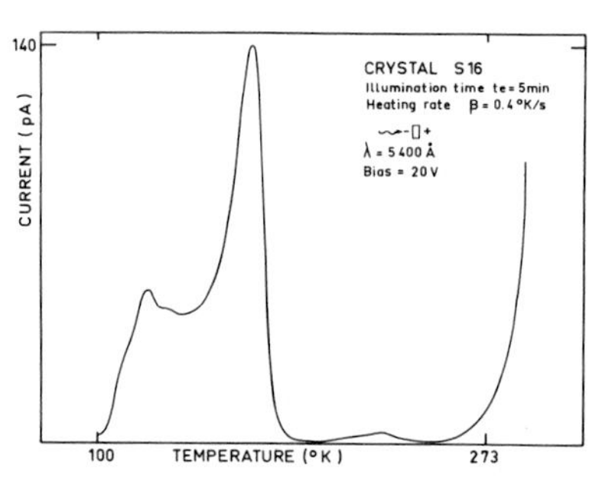

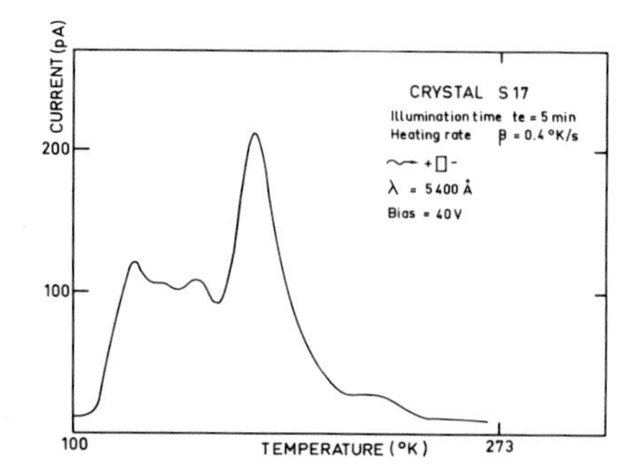

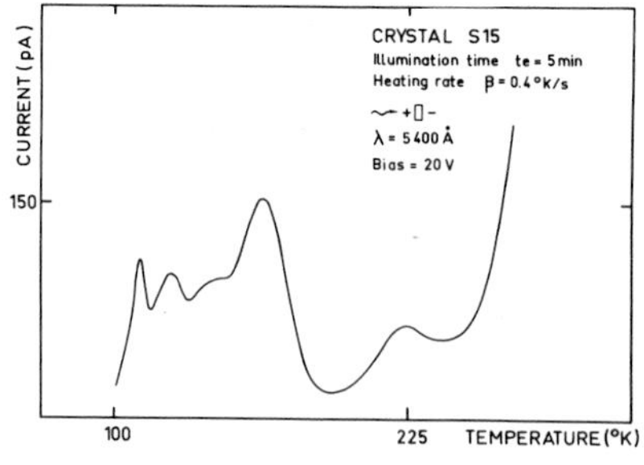

Figs. 5-7 : Typical TSC spectra recorded on the various kind of materails we investigated.

In conclusion, an analysis of the peak located at 0.45 eV (173 K) constitutes a good indication of the crystals possibility for preparing spectrometers of good performance and stability in time.

At that point of our study it cannot be indicated if that level plays the effective role on polarization, since the concentrations of defects at various energetic positions in the gap are not independent, but strongly correlated.

That the higher stoichiometric crystals give the best performance is not surprising, in principle, but we proved it here.

REFERENCES

1. I. F. NICOLAU, G. ROLAND, M. F. ARMAND, J. C. MENARD, Stoichiometry deviation in -HgI_2 to be published, CEA-CENG, LETI/CRM 85X, 38041 GRENOBLE CEDEX FRANCE
2. G. DISHON, M. SCHIEBER, L. BEN-DOR, L. HALITZ, Mat. Res. Bull. 16, 565, (1981).
3. C. SCHARAGER, A. TADJINE, M. TOULEMONDE, J. J. GROB, P. SIFFERT Proceeding of Nuclear Physics, 7th Divisional Conference Nuclear Physics Methods in Material Research, Darmstadt (F. R. G.)(1980)
4. H. SCHOLTZ, Philips Technical Review, 10, 317 (1967)
5. Workshop on HgI_2, University of Strasbourg, Informal Proceeding, edited by P. SIFFERT. (1975).
6. R. H. BUBE, Phys. Rev., 106, 703 (1957).
7. C. MANFREDOTTI, R. MURRI, E. PEPE and D. SEMISA, Phys. Stat. Sol. (a) 20, 477 (1973)
   C. MANFREDOTTI, R. MURRI, A. QUIRINI and L. VASANELLI, Phys. Stat. Sol. (a) 38, 685 (1976)
8. J. P. PONPON, R. STUCK, P. SIFFERT and C. SCHWAB Nucl. Instr. Meth. 119, 194 (1974).
9. U. GELBART, Y. YACOBY and I. BEINGLASS and A. HOLZER IEEE Trans. on Nucl. Sci. NS 24, 135, (1977)
10. R. C. WHITED and L. VAN DEN BERG, IEEE Transactions on Nucl. Science, NS 24, 165 (1977)
11. S. B. HYDER, J. Appl. Phys., 48, 313 (1977)
12. C. DE BLASI, S. GALASSINI, C. MANFREDOTTI, G. MICOCCI, L. RUGGIERO and A. TEPORE, Nucl. Instr. and Methods, 150, 103 (1978)
13. J. SAURA, Thesis Universidad de Cuyo, Argentina (1972).
14. T. MOHAMED BRAHIM , Internal. Report CEA (France) SES/INTERNE SAI/79-190.
15. G. F. J. GARLICK and A. F. GIBSON, Proc. Phys. Soc., 60, 574 (1948).
16. R. STUCK, J. C. MULLER, J. P. PONPON, C. SCHARAGER, C. SCHWAB, P. SIFFERT, Workshop on HgI_2, University Strasbourg, informal Proc. edited by P. SIFFERT. (1975)
17. L. W. ANDREWS, J. American Chem. Soc. 25, 756 (1903)
18. J. ROSIN, "Reagent Chemicals and Standards" 5th Ed. Editor Van Nostram and Co - New York (1943).
19. G. S. JAMESON, Amercian J. Sci., 33, 349 (1912).
20. G. S. JAMESON "Volumetric Iodate Methods", The chemical Catalogue Co - New York (1926).
21. I. M. KOLTHOFF, J. P. ELWING, "Treatrise on chemical chemistry" Interscience Publishers -New York , Part II, 3, 308 (1961).
22. G. SCHWARZENBACH and H. FLASCHKA, "Die komplexometrische Titration"

23. PERRIN, Chemical Analysis vol. 33. "Masking and Demasking of Chemical reactions" Wiley, Interscience - New York. (1970).
24. I.M. KOLTHOFF and J.P. ELWING, "Treatrise on chemical Chemistry" Interscience Publishers -New York PartI, 1, 588 (1961).

TABLE IV

Behaviour of the various crystals to polarization

Crystals	Behaviour
S01 S02 S03 S14	Fast polarization and very poor energy resolut.
S16 S17 S18	No polarization effect and very good energy resolution for X- or Y - rays below 60 keV
S11 S12 S15	Very sensitive to surface preparation conditions Slow polarization effect, good energy resolution up to ^{57}Co Y- rays.

OHMIC AND RECTIFYING CONTACTS ON HIGH RESISTIVITY P-TYPE CADMIUM TELLURIDE

A.MUSA,J.P.PONPON,M.HAGE-ALI
CENTRE DE RECHERCHES NUCLEAIRES ,Laboratoire PHASE
67037 STRASBOURG-Cedex FRANCE

ABSTRACT

Ohmic and rectifying contacts on high resistivity etched P-type cadmium telluride have been studied in order to produce diode structures.For this,we have first investigated the properties of gold contacts obtained by chemical reactions of CdTe dippedin gold chloride.Both electrical characterization and structure have been analyzed as a function of the experimental conditions of the contact deposition.The results can be interpreted in terms of a current flow enhanced by tunnelling through the Au-CdTe junction and related to the structure of the interface a few tens of nanometer below the gold contact.
In addition,several rectifying contacts have been investigated , in order to achieve a structure having low leakage current.

INTRODUCTION

The formation of ohmic and rectifying contacts is highly desirable for a semiconductor nuclear radiation detector, as it allows the device to work in the depletion mode.On the other hand it is known that making a good ohmic contact on P-type CdTe either of low or high resistivity is not easy.In addition,the formation of a blocking contact with a barrier high enough to produce a low leakage current and a high breakdown voltage is difficult to achieve on chemically etched CdTe surfaces.For these reasons,very often , mechanically polished or lapped surfaces are used to fabricate CdTe nuclear radiation detectors.However, the advantages related to the use of heavily damaged surfaces (low leakage current,high biasing voltage) are compensated by disatvantages resulting from the presence of high resistive layers which can be several microns thick (1).This leads to a modification of the electric field in the detector (2) and can contribute to the polarization effect (2,3). In addition,the voltage drop in these layers can noticeably reduce the effective voltage applied in the sensitive volume of these counters.In order to

Mat. Res. Soc. Symp. Proc. Vol. 16 (1983)

overcome these difficulties,we have tried to investigate diodes on P-type etched CdTe by studying the formation of a convenient ohmic contact on high resistivity material and by reviewing various methods giving rectifying contacts.Ohmic contacts were obtained by electroless deposition from gold chloride (4).Their properties have been investigated by using both medium (10E2 -10E3 Ω.cm) and high resistivity (10E6-10E9 Ω.cm) P-type CdTe. Electrical measurements were used to determine their specific contact resistance and their structure was analyzed by means of SIMS and RBS experiments.In the case of rectifying contacts ,the electrical characteristics (leakage current and breakdown voltage) and the detection properties (resolution in energy , polarization) were studied for structures such as M-S,M-I-S, P-N junctions as well as heterojunctions.

GOLD CHLORIDE CONTACT ON CdTe

EXPERIMENTAL

SAMPLE PREPARATION:single crystalline or large grain polycristalline CdTe grown by the THM method has been used.After cutting slices of 2 mm in thickness and lapping by means of abrasive powders,the samples have been chemically etched in an aged bromine in methanol solution with 10 % bromine in order to remove the damage created by the mechanical treatments. For the electrical measurements,contacts of about 2 mm in diameter have been realized by the reaction of a drop of gold chloride solution on the CdTe surface.Symetrical contacts have been deposited on opposites faces on the high resistivity samples , while on the low resistivity material electrolytic deposition of copper on one face was used.This process typically gives a specific contact resistance of about 2 $\Omega.cm^2$ on 300 Ω.cm etched materials. For the SIMS and RBS measurements ,large area contacts have been made simply by dipping the samples in the gold solution , in order to obtain a good uniformity.All the measurements have been performed as a function of the time of reaction of gold chloride with cadmium telluride, from one minute to several hours at room temperature.

Electrical measurements: the influence of the surface preparation on the specific contact resistance ρ_c has been first investigated on 1300 Ω.cm samples with lapped ,polished and etched surfaces. The results are reported on Table I and show the influence on ρ_c of a damage layer on the semiconductor surface.

TABLE I

Influence of the surface preparation on the specific contact resistance of electroless gold contact on 1300 Ω.cm P-type CdTe

Surface preparation	ρ_c $(\Omega.cm^2)$
chemically etched	1.3×10^3
polished	10^4
lapped	2×10^4

In addition , a clear relation was found between the specific contact resistance and the bulk resistivity ρ as ρ_c roughly increases linearly with ρ .

On high resistivity material, a strong influence of the time of reaction on ρ_c could be observed qualitatively. In fig. 1 we have plotted the total resistance of symetrical contacts reduced by the contribution from the bulk.

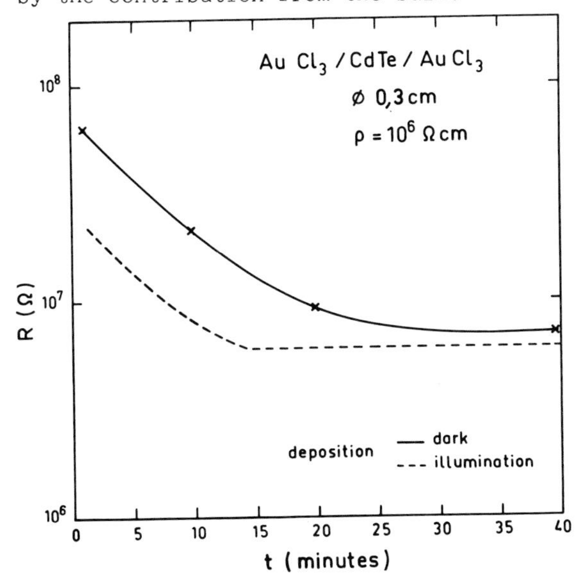

Fig1: Evolution of the total resistance (bulk+contact) as a function of time of reaction of CdTe with $AuCl_3$.

In order to avoid errors due to the lack of accuracy on the sample resistivity ,always the same sample was used for all the measurements. The total resistance clearly decreases with increasing time of reaction , and a saturation is observed after about 40 min. in the dark, giving a resistance about ten times lower than after an 1 min. reaction time. Illumination of the device during the process reduces the saturation time by a factor three to four. Structure analysis: secondary ion mass spectroscopy (SIMS) and Rutherford backscattering (RBS) have been combined to analyze the structure of the gold chloride contacts on CdTe. In fig. 2,3 and 4 are reported SIMS profiles corresponding to a reference sample (100 nm evaporated gold film on CdTe), a sample after 1 min. reaction in $AuCl_3$ an other after 40 min. of reaction, respectively. As expected, the reference sample shows a sharp decrease of the gold distribution at the interface ,which corresponds to a sharp increase of the Cd and Te lines. In contrast, the SIMS results are fully different for samples dipped in gold chloride, especially for the longer time of reaction

In fig. 4 for example, the distribution profiles can be separated ,within the limit of the SIMS edge effect, in four parts: the first corresponds to the surface region and mainly is made of a gold layer which is cadmium and tellurium rich. The second represents the interface and shows an increase in the concentration

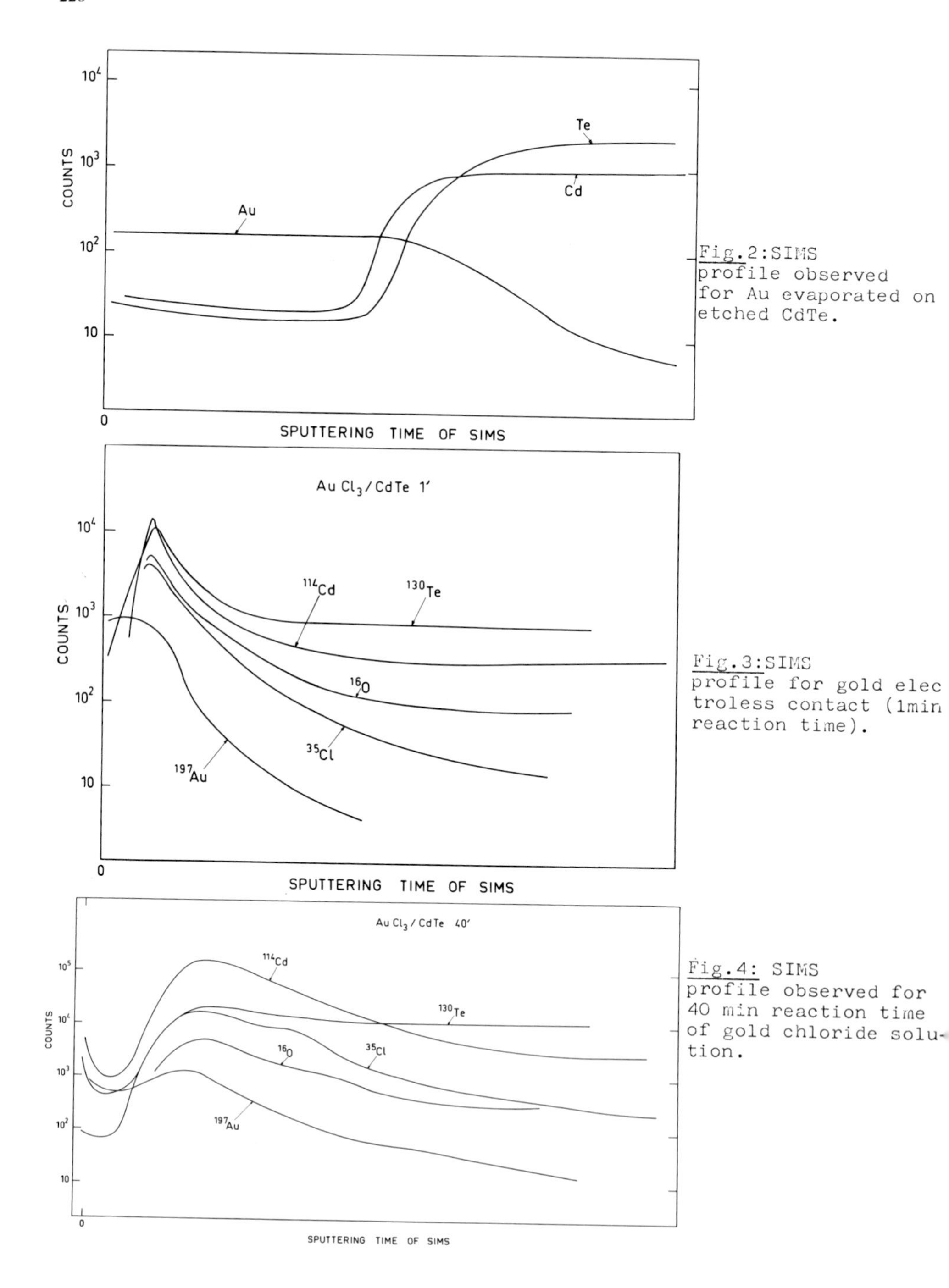

Fig.2: SIMS profile observed for Au evaporated on etched CdTe.

Fig.3: SIMS profile for gold electroless contact (1min reaction time).

Fig.4: SIMS profile observed for 40 min reaction time of gold chloride solution.

of Cd,Te,O and Cl.in this region, the cadmium yield is enhanced by the high concentration of chlorine.The third region beyond the interface is made of CdTe in which O and Cl are present and may be also Au;however,this latter is difficult to estimate as the profiles can be somewhat modified by edge effects during the SIMS analysis.Deeper in the material, a typical bulk CdTe spectrum is found , which is similar to that observed in the reference sample.

Rutherford backscattering experiments have been performed by using 3 MeV $^4He^+$ ions and high resistivity samples.Generally, normal incidence of the ion beam with respect to the surface was used and the backscattered particles were detected at an angle of 13 ° off the normal in front of the sample.For the thinner gold layers (below 30 nm) the impinging angle was 50° and the detection angle 63°,in order to improve the depth resolution.In fig. 5 are plotted the energy distribution of the backscattered particles for samples dipped for 1, 20 and 40 min.,respectively.

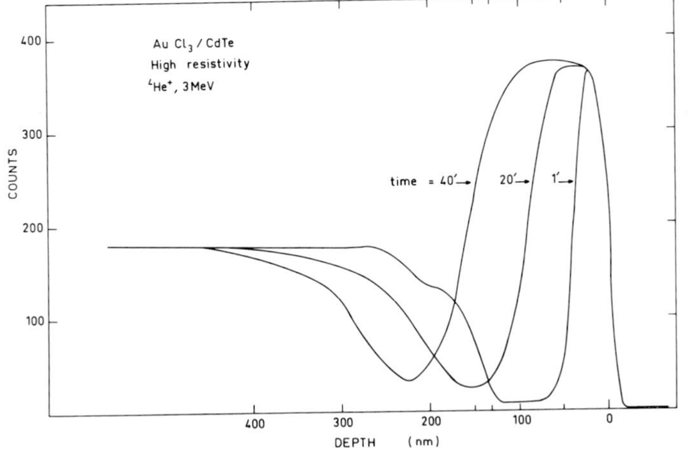

Fig.5 :RBS spectrum for gold electroless contacts for 1 ,20 and 40 min. reaction time.

It is clear from this figure that the total amount of gold deposited on CdTe ,which is proportional to the area of the gold distribution ,increases with the time of reaction.The tail of the gold distribution at low energy side of the spectrum can be interpreted in terms of gold penetration into the material.However, a contribution on it of a non uniform gold thickness due to surface roughness cannot be definitively ruled out , altough the preparation of the samples has been done to avoid this problem.The gold distribution can be separated into two components :the first one , which corresponds to the depth of the plateau ,represents the thickness X_o of the gold film deposited on the surface.Comparison with a reference sample shows that in this region gold deposited by electroless has the same composition as an evaporated gold film. The evolution of X_o with the time of reaction has been studied for times of reaction up to 3 hours and could be expressed by :

$$X_o = 3.61 \times t^{0.42} , \quad (1)$$

with X_o expressed in nm. and t in sec.

The tail which is in the second component of the distribution can then be considered as a penetration of gold into the bulk and be tentatively expressed like a diffusion by an exponential law :

$$N(x) = N_o \exp -(X - X_o)/L , \qquad (2)$$

where N_o represents the gold concentration in the surface layer and X the depth measured from the gold surface ,L is a parameter which characterizes the diffusion and which represents the amount of diffused gold . The evolution of L with time of reaction was found experimentally to vary following the relation :

$$L = 7.45 x 10^2 \times t^{0.51} \qquad (3)$$

where L is expressed in nm.
Finally, it can be noticed that the gold electroless process is associated with a cadmium removing from the CdTe surface, in agreement with other measurements (5).

INTERPRETATION

In addition to the model proposed years ago by DE NOBEL (4), considering a precipitation of gold on a tellurium layer,associated with cadmium removal, our experiments suggest a model of gold diffusion into the bulk CdTe ,which determines the mode of current flow and ,therefore, the specific contact resistance. .Gold concentrations in excess of 1x10 E 17 cm E-3 can be found beyond the interface for times longer than 20 to 40 min.;therefore, a high acceptor concentration exists in such a layer.This doping level even could be enhanced by cadmium vacancies.Then tunnelling between this region and the metal contact is possible, allowing the contact to work according to the model proposed by POPOVIC (7) with a low specific resistance .The influence of tunnelling could be verified on lower resistivity samples as the barrier height which can be associated with the specific contact resistance (0.54 eV) was much lower than the lowest barrier height which can be made on P-type CdTe (i.e. 0.64 eV). This indicates that pure thermoionic emission is not the only transport mechanism in this contact.In any case, despite the high supposed gold concentration atoms introduced in CdTe and due to their low doping efficiency,as noticed by AKUTAGAWA (8), the specific contact resistance which results from a gold electroless contact is much higher than that which could be obtained by electrolytic copper deposition for example.Thus, if such process cannot be of use for low resistivity material,it is , however, good enough to give ohmic contacts on high resistivity devices ,provided care has been taken during electroless deposition to insure incorporation of gold concentrations large enough into a close surface layer.

RECTIFYING CONTACTS ON ETCHED PTYPE CdTe

SAMPLE PREPARATION: crystals grown by THM method have also been used here.They were generally 10 E6 to 10 E 9 Ω .cm in resistivity and most of them gave detectors showing polarization effects when contacted with aluminium on lapped surfaces.Samples 300 microns and up to 1 mm in thickness have been etched as above and care has been taken to remove the damaged layers on the surfaces.The ohmic contacts have been prepared as indicated above for 40 min. reaction time of the gold chloride solution.The recti fying contacts were typically 3-4 mm in diameter and several methods have been used for their preparation.They will be listed below and we shall indicate for each of them the main I-V and detector's characteristics that were obtained.The energy resolution have been measured at room temperature for ^{57}Co (122 keV). These materials when processed in the conventional way (aluminium contacts) gave 7-10 keV energy resolutions.

METAL - SEMICONDUCTOR CONTACTS

Due to the low barrier heights that can be achieved on P-type etched CdTe,low breakdown voltages can be expected for such structures .Detectors have been made by using either evaporated indium contacts or aquadag.The mean electric field that could be applied in these counters was no more than 1.5 kV/cm.However, good detection properties were obtained (7 keV resolution) especially very low polarization was observed even with crystals giving high polarization on lapped deveices.

METAL-INSULATOR-SEMICONDUCTOR CONTACT

In order to improve the reverse characteristics MIS structures made of aluminium evaporation over 700 nm. thick silicon oxide deposited on CdTe were used.The mean electric field in the detectors was increased by a factor three,however, poor detection properties resulted due to excess noise in the oxide layer .As already observed (9) the polarization was strongly reduced.

N-P JUNCTIONS

As it has been shown that Indium can diffuse in large amount in CdTe (10),we have tried to prepare P-N junctions by heating up to 160 °C for 30 min. 500 nm. thick In dots evaporated on CdTe surfaces.Good I-V characteristics were obtained and mean electric fields up to 8-9 kV could be applied on the detectors, which did not show polarization effetcs ; the energy resolution was about 7 keV.Attempt to induce In diffusion by pulsed laser induced diffusion were not successfull as probably due to high damaging of the space charge region , since the charge collection efficiency was drastically reduced.

HETEROJUNCTIONS

One way to produce a high band bending on CdTe is to use a heterojunction (11).In orderto evaluate its advantage when used in a nuclear detector ,we have evaporated indium tin oxide layers 400 nm. thick on CdTe.On low resistivity materials the barrier height was found to be 0.8-0.96 eV ,strongly depending on evaporation conditions .On high resistivity samples ,low leakage current was obtained and a rather high electric field could be applied (4 kV/cm)Some polarization effects ,however, were observed altough they were lower than in the corresponding Al/lapped detectors.Energy resolution was about 10 keV.

CONCLUSION

With the help of a few results we have shown that the use of both an appropriate ohmic contact and a rectifying one with band bending high enough,can be used to prepare nuclear radiation detectors on etched cadmium telluride surfaces.,then avoiding the voltage drop in the heavily damaged layer and reducing that fraction of polarization related to that surface layer.Gold electroless deposition is able to give contacts good enough on high resistivity P-type material ,while it can be expected that P-N junctions and even heterojunctions are promising structures to built detectors on semi-insulating CdTe with etched surfaces.

ACKNOWLEDGMENTS
Many thanks are due to J.M.KOEBEL who prepared the various crystals used in this work and Mrs RIT and AMANN who prepared the samples.
The SIMS measurements have been performed by R.STUCK and J.J.GROB contributed to the RBS experiments.

REFERENCES

1. M.HAGE-ALI,R.STUCK,C.SCHARAGER,P.SIFFERT IEEE Trans.Nucl.Sci. NS 26 (1979)281
2. P.G.KASHERININOV Sov.Phys.Semicond. 15 (1981) 1099
3. R.HOFSTADTER Nucleonics 2 (1949)29
4. D.DENOBEL Philips Res.Rep. 14 (1959) 361
5. A.M.MANCINI,A.QUIRINI,L.VASANELLI,E.PERILLO,E.ROSATO,G.SPADACCINI J.Appl.Phys. 53 (1982)5785.
   idem Thin Solid Films 89(1982)407
6. J.P.PONPON,M.SARAPHY,E.BUTTUNG,P.SIFFERT Phys.Stat.Sol. (a) 59(1980)259.
7. R.S.POPOVIC Sol.State Electr. 21(1978)1133
8. V.AKUTAGAWA,D.TURNBULL,W.K.CHU,J.W.MAYER,J.Phys.Chem.Sol. 36 (1975) 521
9. P.SIFFERT,M.HAGE ALI,R.STUCK,A.CORNET,Rev.Phys.Appl.12(1977) 335.
10. R.E.BRAITHWAITE,C.G.SCOTT,J.B.MULLIN,Sol.Stat.Electr. 23(1980) 1091
11. A.L.FAHRENBRUCH,F.BUCH,K.MITCHELL,R.H.BUBE,IIth IEEE Photovoltaïc Specialists Conference 1975 p.490·

A PRELIMINARY OBSERVATION ON $AgGaSe_2$ AS A NUCLEAR RADIATION DETECTOR

EIJI SAKAI, HIROMICHI HORINAKA*, HAJIMU SONOMURA* AND TAKESHI MIYAUCHI*
Japan Atomic Energy Research Institute, Tokaimura, Nakagun, Ibarakiken, Japan 319-11

ABSTRACT

An about 10^8 ohm-cm $AgGaSe_2$ crystal of 0.5 mm x 4 mm x 4 mm was polished and contacts were made by evaporating 130 $\mu g/cm^2$ gold of 3 mm diameter on the two faces of the crystal. The detector was tested using 5.5 MeV alpha-particles at room temperature. Noise increased above an applied bias voltage of 80 V. For + 50 V applied on the electrode opposite to the particle incident electrode, i.e., for electron traversal mode, the preamplifier output pulses showed a risetime of 20 µs and an amplitude of about one-tenth of that obtained from a silicon surface-barrier detector whereas the silicon detector showed a risetime of 0.07 µs. For - 50 V applied on the same electrode, i.e., for hole traversal mode, no pulses were observed.

INTRODUCTION

GaAs, CdTe, HgI_2 [1], and very recently CdSe [2] have been tried as room temperature large atomic number semiconductor detectors and showed reasonable results as X- and gamma-ray detectors with some difficulties to be solved. Although it is certainly necessary to seek for new materials suitable for radiation detection, not many works are being carried out in this direction. We report here the results of a preliminary observation on a ternary compound $AgGaSe_2$ as a radiation detector. $AgGaSe_2$ is being studied as a material for non-linear optical application and has a combined atomic number Z = 47-31-34, a density $\rho = 5.71\ g/cm^3$ and a band gap $E_G = 1.80$ eV (300K) whereas no data was found on carrier mobilities, energies per electron-hole pair, etc. It is interesting to investigate carrier transport properties using alpha-particle bombardment from the view point of both radiation detection and solid-state physics although difficult problems such as stoichiometry are anticipated to be more severe in ternary compounds than in binary compounds.

DETECTOR

$AgGaSe_2$ crystals (melting point = 850 °C) were grown by Bridgman method in a doubly-sealed quartz ampoule which contained Ag, Ga and Se metals of a purity of five-nines. The crystals of 10^8 ohm-cm resistivity were cut with a wire saw, flattened on a whetstone and polished on a rotating sheet of velvet. A single crystal of 0.5 mm x 4 mm x 4 mm was ultrasonically cleaned in ethyl alcohol, a gold film of 130 $\mu g/cm^2$ was vacuum-evaporated on the areas of 3 mm diameter as a front and a back electrode and the back electrode was directly mounted on the center pin of a BNC-R type connector using silver conducting cement. The front electrode was grounded to an aluminum case using a thin gold wire and conducting cement as shown in Fig.1 (next page).

*College of Engineering, University of Osaka Prefecture, 4-804, Mozuumemachi, Sakaishi, Osakafu, Japan 591.

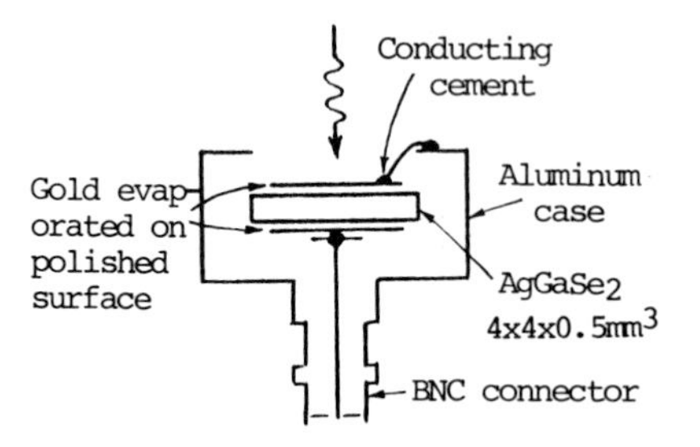

Fig. 1. Detector construction.

MEASUREMENT APPARATUS

Figure 2 shows a schematic diagram of the electronics used in the detector performance measurements. The detector was placed in a vacuum chamber with an Am-241 alpha-particle source. The signals from the detector were amplified with a charge-sensitive preamplifier JAERI 125 of ac-coupled configuration and a main amplifier Canberra Industries (CI) 2001, and analyzed with a 4096-channel pulse height analyzer CI 8100. The detector bias voltage was supplied from a high voltage power supply CI 3005. A pulse generator ORTEC 419 was used for monitoring the stability and the noise of the electronics system. The output pulses from the preamplifier and the main amplifier were monitored using a oscilloscope Tektronix 465.

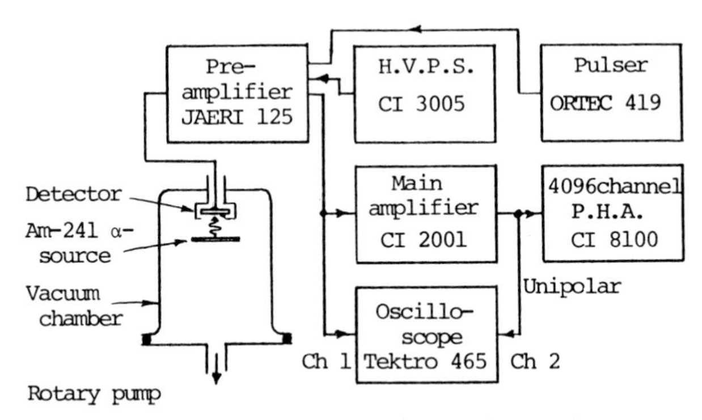

Fig. 2. Schematic diagram of electronics used.

RESULTS

All the measurements were made at room temperature. The $AgGaSe_2$ detector exhibited an increase in noise at bias voltages above |± 80 V|, which corresponded to an electric field of 1600 V/cm (= 80V / 0.05cm), on the back electrode of the detector. The detector showed output pulses only for positive bias voltages, but no pulses for negative voltages; this means that electrons liberated by the alpha-particles could be collected while holes could not be collected.

Figure 3 (c) shows the pulse shapes observed at the preamplifier output from the $AgGaSe_2$ detector with + 50 V bias, i.e., in electron traversal operation mode, which is compared with those shown in (a) and (b) from a silicon surface-barrier detector (ORTEC BA-19-150-100 s/n 14-709D, + 120 V) when both the detectors were irradiated by 5.5 MeV alpha-particles from an Am-241 source. The pulses from the $AgGaSe_2$ detector showed a risetime of 20 μs in contrast to 0.07 μs risetime from the silicon detector. These slow-rising pulse shapes observed in the $AgGaSe_2$ detector are caused by a severe trapping of electrons and the deduction of the electron mobilities is difficult unless more higher bias voltages become applicable. Figure 4 (next page) shows a comparison of the pulse height distributions obtained from both the detectors using the same gain and the same shaping time of 8 μs of the main amplifier. The pulse heights observed from the $AgGaSe_2$ detector were less than one-tenth of those from the silicon detector.

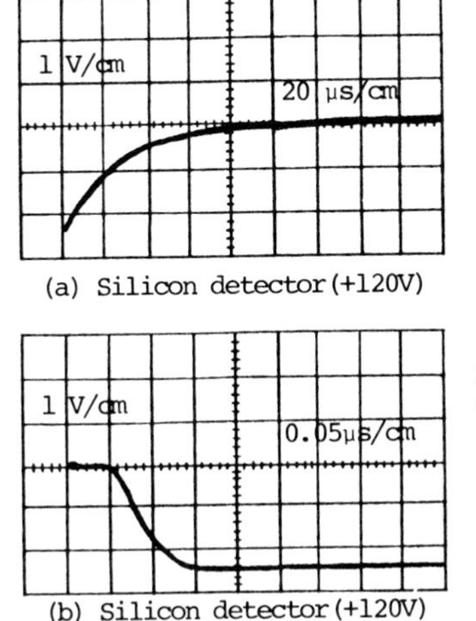

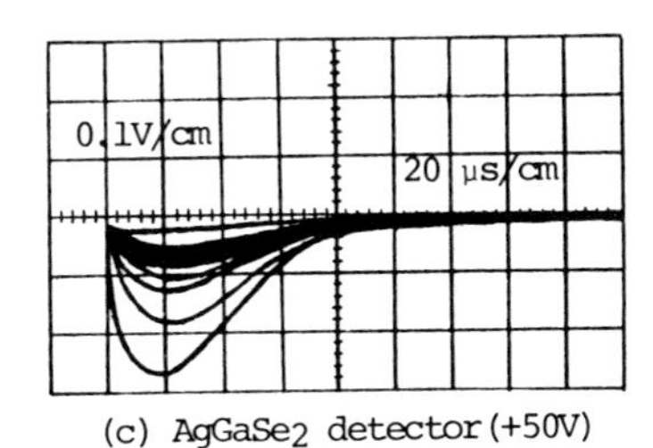

Fig. 3. 5.5 MeV alpha-particle-pulses observed at preamplifier output.
(a) Silicon detector (+120 V)
Vert. = 1 V/cm, Horiz. = 20 μs/cm
(b) Silicon detector (+120 V)
Vert. = 1 V/cm, Horiz. = 0.05 μs/cm
(c) $AgGaSe_2$ detector (+ 50 V)
Vert. = 0.1 V/cm, Horiz. = 20 μs/cm.

CONCLUSION

The first trial of a $AgGaSe_2$ crystal as a nuclear radiation detector was made in which 5.5 MeV alpha-particle pulses with slow risetimes and insufficient pulse heights were observed only for electron traversal mode whereas no pulses were observed in hole traversal mode. Both types of carriers are very severely trapped in the present $AgGaSe_2$ crystal.

REFERENCES

1. For example, E. Sakai, Nucl. Instr. and Meth. 196, 121 (1982).

2. A. Burger, I. Shilo and M. Schieber, IEEE 1982 Nucl. Sci. Symp., 1G10, Washington, D. C. (October 1982). To be published in IEEE Trans. Nucl. Sci. NS-30, No.1 (February 1983).

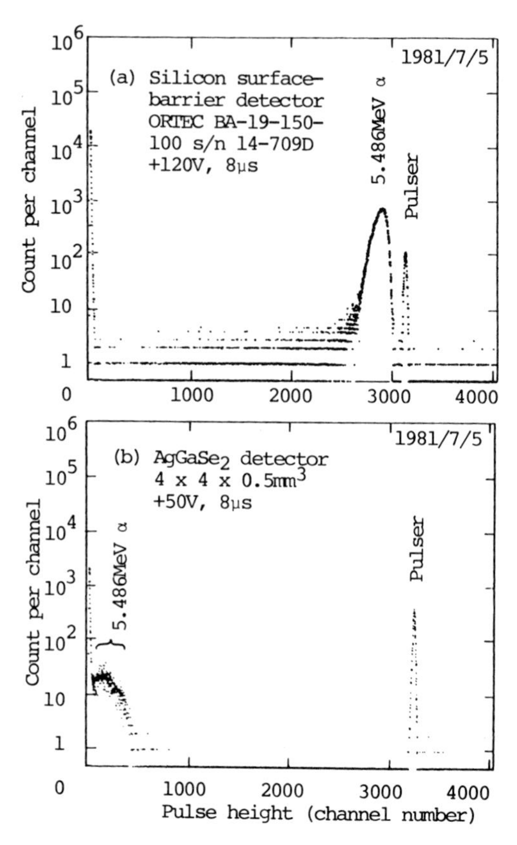

Fig. 4. Pulse height distributions of Am-241 alpha-particles obtained from (a) silicon surface-barrier detector (+ 120 V) and (b) $AgGaSe_2$ detector (+ 50V). The gain and the shaping time (8 μs) of the main amplifier were the same in both cases.

Author Index

Subject Index